FIRE SERVICE AND THE LAW

Second Edition

by Timothy Callahan

and Charles W. Bahme

National Fire Protection Association
Quincy, Massachusetts

J2-2210201

Project Manager: Gene A. Moulton
Editor: Marion Cole
Designer: Sam Stevens
Illustrator: Susan LinVille
Cover: Frank Lucas
Composition: Sharon L. Summers
Peggy Travers
Production Coordinators: Elizabeth K. Carmichael
Debra A. Rose

NFPA Catalog No. FSP-3B
ISBN 0-87765-338-O
Library of Congress Catalog Card No. 87-61174
Printed in the United States of America.

Contents

Foreword

Many changes have occurred during the 40-plus years since the original publication of Chief Bahme's dissertation, "Legal Problems of Municipal Fire Departments." That work was the forerunner of his other textbooks on fire service law and of the First Edition of this book. Since the First Edition's publication in 1976, the scope of fire department responsibility has greatly expanded. The body of law and the amount of litigation on the subject also have increased several fold.

During the last decade, numerous court decisions, administrative regulations, ordinances, and statutes have affected the legal position of all members of the fire service. Changes in legal doctrine and standards for fire service performance have placed greater responsibility on every member of the fire department. Such increases have led to greater personal and department exposure to liability. To inform the fire service of these changes and their consequences is the aim of this edition.

Chief Bahme's unique historic perspective and familiarity with both the legal and the fire service environment make his contributions on the subject particularly incisive. NFPA and the fire service are fortunate to have received the benefit of his work.

With this edition, NFPA welcomes a new author who actually is an old friend. It is our good fortune to have persuaded Tim Callahan—a lawyer, a fire protection engineer, and a former NFPA staff member—to prepare the revision of this book now that Chief Bahme is retiring from textbook writing. I know that the final result of their collaboration on the Second Edition is a valuable reference book for every fire service member.

Robert W. Grant, President
National Fire Protection Association

Preface

This work has been prepared as a practical, comprehensive, and up-to-date resource for all members of the fire service. It is designed not merely to give instruction in the law and the legal practices affecting rendition of fire department services, but also to identify areas of particular fire service concern that may warrant attention.

In preparing this edition, I have relied upon Chief Bahme's prior works to identify areas of historic fire service interest and concern. To this were added subjects of current or particular controversy as evidenced by writings in fire department and fire service publications. The author welcomes all suggestions on subjects covered—especially new cases or items for inclusion in subsequent editions.

While every effort has been made to check the accuracy of quotations and cases cited, no warranty is expressed or implied, nor is specific counsel intended by the contents of this work. Competent local counsel should be sought in each instance.

It is my hope that, in reading this work, the fire service will gain an understanding of how it operates within the bounds of our nation's laws and how, through these laws, the citizenry's injury and suffering due to the ravages of fire can be diminished.

Timothy ("Tim") Callahan
February 1987

Acknowledgments

First, I wish to express my appreciation to Sandra Ries, Golden Gate University School of Law Class of 1987, for her update on First Edition citations and her research on the subjects of random drug testing and fire fighter safety. Her suggestions served as the basis for presentation of these subjects.

Grateful acknowledgment for suggested changes and additions to the First Edition is extended to Gordon Routley, Neil Schreck, Gary Smith, and all my former students at Cogswell College's Open Learning Fire Service Program.

To Dr. John L. Bryan and the late Harry E. Bigglestone, whose insight into public fire protection and support for my study of the law were of immeasurable help, go my sincere appreciation.

To Marion Cole and Gene Moulton, Editorial Department at NFPA, go my thanks for their patience and constructive criticism. My appreciation also is extended to Dennis J. Berry, NFPA's Associate General Counsel, for his invaluable suggestions.

Finally, to my family—wife Kathy and sons Brian and Matthew, for their patience and understanding throughout the writing period, and my father, Stanton Callahan, for his layman's critique of the manuscript—go my heartfelt thanks.

T.C.

The Authors

Timothy ("Tim") Callahan is President of Fire Protection Consultants Inc., a fire protection engineering and consulting firm based in Walnut Creek, California. His 14 years of fire protection-related experience include positions as fire fighter and apparatus operator, fire company administrative officer, and field service engineer; as a Public Fire Protection Specialist with the National Fire Protection Association; and as chief of fire protection services for United States Army forces in Europe.

Previous publications by Tim include articles in *Fire Command* and the United Kingdom periodical *Fire Prevention*, as well as within the National Fire Academy's Open Learning Fire Service Program for the curriculum on "Incendiary Fire Investigation."

Tim received a Bachelor of Science degree in fire protection engineering from the University of Maryland; a Master of Science degree in business administration from Boston University; and a Doctor of Jurisprudence degree from Golden Gate University. He is admitted to the bar and a licensed fire protection engineer in California.

Charles W. ("Chuck") Bahme is the author of many fire service publications including *Fireman's Law Book*, predecessor to *Fire Service and the Law*. In 30 years of service with the City of Los Angeles Fire Department, he advanced through the ranks from recruit to Deputy Fire Chief, his position at the time of retirement. During that period he augmented his Bachelor of Arts degree from the University of California at Los Angeles (UCLA) with a Juris Doctor degree from Southwestern University. Through the same years he also served with the U. S. Navy: six years of active duty during World War II and the Korean conflict, and 28 years with the Naval Reserve from which he retired with the rank of Captain.

Chief Bahme is an attorney at law with admission to practice in California, before Federal district courts, the Supreme Court of the United States, and the highest court of military appeals. He has taught courses in fire protection engineering at UCLA and in fire administration at the University of Southern California, and has lectured at various state, national, and international conferences. He has served in Europe and the Far East for the U. S. Department of Defense and the State Department.

In recent years, Chief Bahme has been teaching a course in "Political and Legal Foundations of Fire Protection" in Cogswell College's Open Learning Fire Service Program. He also has served NFPA as the Fire Extinguishing Systems Specialist, as its first West Coast Representative, and as chairman of the Sectional Technical Committee on Extinguishment of Chemical Fires.

Now a resident of Lakeport in Lake County, California, Chief Bahme has been a disaster control advisor to the local chapter of the American Red Cross and chairman of the board of directors of his home fire district.

The following abbreviations are used in this work, especially in the footnotes:

Ibid.—"In the same place"; exactly the same reference as the one immediately before this one.

Op. cit.—"In the work cited"; refers to the publication by this author which was cited earlier in the chapter.

Supra—"Above"; refers to a legal case or footnote reference cited earlier in the chapter.

All legal cases cited are listed alphabetically, by chapter, in the Table of Cases that follows Chapter 14. Use of these case references will facilitate retrieval of published decisions from a law library.

The reader's attention also is called to the Glossary of Legal Terms that follows the Table of Cases at the back of this book.

Chapter One

Introduction to the Law

The Meaning of Law

The word "law" is an all-encompassing term referring to the body of rules of conduct prescribed by Congress, state legislatures, the courts, or common community practice applicable to everyone. "Law is used in a generic sense, as meaning the rules of action or conduct duly prescribed by controlling authority, and having binding legal force," as defined in *U.S. Fidelity and Guarantee Co. v. Guenther*. Viewed in this light, "law" includes both written and unwritten law, civil and criminal law, procedural and administrative law, and understandings of conduct not explicitly stated but commonly observed.

Every law may be said to consist of several parts: one, declaratory, whereby the rights to be observed and the wrongs to be avoided are clearly defined and laid down; another, directory, whereby the subject is instructed to observe those rights and abstain from the commission of those wrongs; a third, remedial, whereby a method to recover a person's private rights or redress private wrongs is pointed out; to which may be added a fourth part, usually termed the sanction or vindicatory branch of the law, which sets forth what evil or penalty shall be incurred by those who commit any public wrong and transgress or neglect their duty. (*State v. Sawn*) Thus, the "law" is a continuously evolving, all-encompassing set of rules of individual conduct designed to preserve and protect the interrelationships of people and the environment.

Sources of Law

How does someone determine definitively what is the law and in what book to find the answer? The reply is that referral to a single document or compendium will not always provide the answer. Usually, it is not enough to merely refer to the statutes of a particular state to ascertain the law on a certain point; rather it is necessary to refer to all sources of the law affecting the subject in a given state. This includes the Constitution of the United States and of that particular state; laws passed by Congress and that state's legislature; the regulations of the Federal government and that state's administrative agencies; ordinances of the affected county or city; decisions of their courts; and other sources. Clearly, no single source will definitively outline the law; on the contrary, reference to a compilation of documents and other written sources is required.

Constitutional Law

A constitution contains the fundamental, organic law or principles of a nation, state, society, or other organized body of individuals. For example, unlike England which has no written constitution, the basic law of the land in the United States is the U.S. Constitution and the constitutions of the several states and commonwealths. These constitutions are supreme, above all statutory and decisional law rendered by Federal, state, and local entities. Any right guaranteed by a constitution is spoken of as a constitutional right, such as the freedoms of speech and of the press outlined in the First Amendment to the United States Constitution. Such a right cannot be abolished by the Federal government, a state, a city, or any other political subdivision. Most fundamental rights, however, are not considered absolute, but can be abridged in the case of a compelling governmental interest. For example, although everyone has a constitutionally protected right of freedom of speech, no one has a right to yell "fire" in a crowded theater. Because the government's interest in preventing panic is considered compelling, an individual's freedom of speech must be bridled to protect the safety of the public.

Most state constitutions are similar to the Federal Constitution in that no state or local laws and regulations may be in opposition to the Federal Constitution's requirements. In one sense, the charter of a city, granted by a state legislature, is the city's constitution. The charter acts not only as a measure of the grant of power conferred upon the city, but also as a means of testing the validity of the ordinances adopted by it. A city's ordinances must not conflict with its charter provisions, its state constitution, nor the Federal Constitution.

Statutory Law

Laws passed by the Congress and not vetoed by the President are termed "statutory law." Within this classification are laws passed by state legislatures and not vetoed by these states' chief executives. In its broadest sense, statutory law includes all laws or other enactments passed by any legislative body, from Congress down to the local community or district council.

TEN YEARS AFTER THE SIGNING of the Declaration of Independence in 1776, the "founding fathers" of the United States created the U.S. Constitution. Historians have called it "the greatest single document struck off by the hand and mind of man."

Because a piece of legislation may address vastly diverse subject areas, researching the law by reviewing all bills passed by the legislature can be needlessly time-consuming. To solve this problem and to facilitate legal research, "codes" or compendiums of all laws on a particular subject are compiled and published. Thus, for example, a state's health and safety code would contain all laws passed by its legislature, from its first session through today, affecting public safety. A state's penal code would designate all the possible acts prohibited by that state. Although the titles of individual codes vary from state to state, there is no variation in the process of compiling all laws on a subject into sets of codes.

Other codes or compilations of the state law of interest to fire service personnel include a state's government code, which might outline the requirements for formation of a fire district; a vehicle code, which might give registration requirements for fire department vehicles; and a labor code, which could include regulations governing the working hours and conditions for fire fighters.

On the local level, communities frequently pass ordinances regulating the construction and operation of buildings; these would be stated in a community's municipal code. For instance, a community might pass requirements that limit combustible vegetation near residential structures, that direct provision of sprinkler protection in buildings housing hazardous processes, or that place limitations on the storage of hazardous materials in school structures. Although they address diverse subject areas and occupancies, these enactments typically would be compiled into a single reference document or code. Thus, when governments adopt regulations involving fire safety—whether they address issues of electrical wiring, structural design, ventilation, or fire prevention—these regulations generally will be compiled and published in one document or one section of a community's municipal code.

Common Law

The rules ascribed to ordinary custom and usage or the common, everyday activities of people are often referred to as the "common law." As distinguished from statutory law, the common law comprises the body of principles and rules of conduct, relating to the government and security of people and property, which derive their authority solely from usage and custom of time immemorial; or the judgments and decrees of the courts recognizing, affirming, and enforcing such usages and customs.

Although many states have adopted the common law by statute, the common law generally can be determined only by a review of all court decisions on the particular subject in question. Through a court's repeated affirmation of a specific principle, that principle takes on the stature of a law. Therefore, it is possible to predict, by review of all court decisions, how that court will rule when presented with a similar question. Thus, the common law takes its form not by the specific words or proclamations of a legislature, but by the power of judicial precedent. As a result, a court generally will rule in the same way when presented with a similar question in the future.

Summary of Types of Law

In summary, law can be classified as constitutional, statutory, or common law. Because conflict may arise among the three types, as well as questions as to which type of law applies to a particular situation or has precedence, it often is difficult to determine exactly what is the law. Frequently asked questions include: How does this new enactment affect current law? Is it constitutional or does it adversely affect an individual's constitutionally protected rights? In this specific set of circumstances, is the new law applicable, to what extent, and within what framework?

Generally, it is possible—with a fair degree of accuracy and certainty—to identify the proper rule of law for a given situation by studying the constitution, statutes, and decisions of courts on the subject. If after doing this and finding no statute or decision that directly addresses the question or is on point, the next step is to try to predict what a judge or jury might decide under the circumstances. A prediction, or course, can be wrong, and yet be in agreement with all legal precedent, because the law is a constantly evolving set of rules, principles, and understandings. Changes in response to new social needs always are possible.

Court Decisions

Sometimes the decisions of courts—even those of the U.S. Supreme Court itself—are later reversed by a different set of judges or by the same court basing decisions on different reasoning or different public needs. Thus, the decisional law of the country is always subject to change. Although uniform model laws on a number of subjects are widely adopted by the various states, there is no obligation upon the states to adopt the uniform laws. This can result in the statutory law and decisions of the courts in one state being exactly opposite those in another, despite identical facts and circumstances. Sometimes, too, the appellate courts of different districts in the same state do not agree. When this happens, the law on that subject within that state remains uncertain until the matter is appealed for resolution to a higher court within that state.

It also can make a difference whether a case is brought in a Federal or state court, because as independent, sovereign judicial bodies the state courts have different procedural requirements, different rules of evidence, and frequently differing opinions on the same legal question. This is why, when the subject matter permits a choice of courts, the attorney will seek to try the case in the court most likely to hand down a decision favorable to his client.

Administrative Rules and Rule-Making

To clarify, implement, and enforce the various laws enacted by Congress, by state legislatures, and by local community councils, each has created administrative agencies or commissions responsible for providing the on-going technical detail and operational procedures necessary to carry out a new law. Within the bounds of the statutory power granted them, these agencies are given wide discretion regarding development of specific requirements and implementation of procedures for achieving compliance with the law. These administrative bodies can be considered as having quasi-legislative and judicial powers, for not only do they promulgate the technical rules, regulations, and procedures, but frequently they also have broad powers of enforcement. Examples of administrative agencies in government are the Federal Communications Commission, Federal Aviation Administration, Environmental Protection Agency, Internal Revenue Service, and the various State Fire Marshal offices.

The Reasons for Laws

Why do we have laws? A solitary person on a desert island has no need for law. There is no one upon whom to impose any rules, nor is there any commanding authority whose laws must be obeyed. In a society of people, however, there are many reasons why laws are necessary, even though they result in restriction of individual liberty. For example, a person's liberty to swing a fist in the air must stop at the point where the first is likely to hit, or appear to hit, another person's nose. Otherwise, there is likely to be a fight or other breach of the peace. Among the objectives which the law seeks to accomplish are the following:

1. Eliminate friction among people, thus preventing disturbances of the peace. Simply stated, the law helps to maintain public order.
2. Protect the personal rights of individuals, such as freedom from injury, right of assembly, and freedom of speech and of the press. In broad terms, the law protects the individual from the excesses and arbitrary actions of government.
3. Protect the property rights of people, as to both personal belongings and real estate.
4. Establish standards of conduct so that a person may know what course of action is expected under given circumstances, and at the same time may be able to predict what another person will do in the same situation. For example, safe operation of motor vehicles is dependent upon this objective.
5. Provide remedies and means to obtain those remedies in case of unauthorized invasions of personal rights.
6. Provide for the punishment of those who do not respect the rights of others, thereby discouraging antisocial behavior that affects the well-being of the general public.
7. Attain efficiency, harmony, and balance of power among all levels of government.
8. Assure that people have the opportunity to enjoy the minimum decencies and pleasures of life.[1]

To carry out the eight objectives listed above, a vast body of law has been developed. As part of this introductory study of the law, this book will consider in the next two chapters the two principal kinds of action dealt with by the courts: civil actions—actions by an individual to seek redress for grievances; and criminal actions—actions by the government against an individual to punish antisocial behavior.

Review Activities

1. In a one-page narrative, state the principal types of law, provide a definition or outline of each, and indicate, where possible, whether one type would have precedence over another type.

---

[1] Mermin, Samuel, *Law and the Legal System*, Little, Brown and Co., 1973

2. The charter of a city, granted by a state legislature, can be considered a city's constitution or the organic law of that community. Given that it is granted by the state legislature, can that charter contain provisions contrary to that state's constitution? To the Federal Constitution? Why or why not?
3. For what reasons might an unfavorable verdict be appealed to a higher court?
4. With a group of classmates, outline how a state administrative agency is created; list reasons why one would be created; and provide an example of a state administrative agency whose responsibilities affect the delivery of a community's fire protection services.
5. With two or more classmates, discuss the reasons for laws and some of the objectives the law seeks to accomplish. Then give at least five one- or two-sentence examples that illustrate why laws are necessary even though they result in the restriction of personal liberty.
6. In a one-page narrative, state why laws are important to you as a member of the general public.

Chapter Two

Civil Actions

What Is a "Lawsuit"?

A civil action is prosecuted by one party against another for the declaration, enforcement, or protection of a right, or the redress or prevention of a wrong.[1] In fundamental terms, it is an effort by one person, with the assistance of the public courts, to obtain justice from someone else. Distinguished from a criminal action, where the party seeking justice is always the government, a civil action can be brought by the government or any individual seeking protection of some right or redress from some wrong. It usually is possible to tell by looking at the name of a case whether it is a report of a civil action or a criminal action. The title given the lawsuit should make this clear.

To understand the title of a lawsuit, use this example: *John Doe v. Richard Roe*, 214 Cal. 241, 4 P.2d 929 (1931). This means that John Doe sued Richard Roe; v. means versus (against); the lawsuit (case) is set forth in volume 214 of the *California Reports* beginning on page 241; the last part of the citation indicates that the case is also reported in another set of books called the *Pacific Reporter* in volume four of the second edition, beginning at page 929; and the year of the court's decision was 1931. To cite a case means to refer to it, as in this example, or to quote it as presenting authority for a principle of law. The example given here is of a civil case, for one man suing another, and the state is not indicated as a party in the lawsuit.

The parties in a lawsuit are the "plaintiff" and the "defendant." The plaintiff is the one who causes the action to be brought against the other person. In the example, John Doe is the plaintiff—the party alleging some injury caused by the defendant—and Richard Roe is the defendant, being the person de-

[1] California Code of Civil Procedure, Section 30

nying the allegations claimed by the plaintiff. Sometimes the parties in a lawsuit are referred to by other names. For example, if the case is appealed to a higher court, the party who appeals is called the "appellant." It makes no difference whether the party who appeals was the plaintiff or the defendant at the initial trial; that party becomes the appellant in the higher court, appealing to have overruled the decision of the lower court. The party against whom the appeal is filed is called the "respondent," for that party must answer to the appeal and potentially defend the trial court's actions in the appellate court.

The object of a lawsuit usually is to seek damages for some injury caused to the plaintiff by the defendant. Civil actions can seek directives or orders from the court other than the payment of money. For example, a plaintiff could seek a court-enforceable directive that a defendant refrain from performing some act injurious to the plaintiff's person or property. Such an action typically is sought by a Fire Prevention Bureau to mandate that the defendant cease performance of some hazardous act.

Alternatively, a plaintiff could seek a court-enforceable directive mandating that the defendant perform some act of benefit to the plaintiff or the public. An example, again in the fire prevention context, would be a directive against the defendant mandating that his land be cleared of highly combustible vegetation. Other examples of remedies sought from courts would be to recover damages for breach of contract, custody of a child, return of property, and probating a will.

Clearly, the remedial actions that can be sought from a court, commonly termed "remedies," are broad and extend beyond directives for the payment of money.

Distinction Between a Civil and a Criminal Action

Sometimes a civil suit and a criminal prosecution can be brought for the same act. Although a civil action cannot be tried along with a criminal prosecution, a plaintiff can sue in a civil court for damages while the defendant concurrently is being tried for committing the crime which resulted in the plaintiff's injuries. For example, if a man unlawfully hits you—a crime called "battery"—you can sue him for recovery of your losses (that is, lost wages, medical expenses, pain and suffering, etc.). Also, the state can prosecute him for the commission of the crime. Thus, two separate and distinct trials can arise from the same act of the defendant.

The primary objectives of a criminal action are to punish the defendant for his wrongful acts and to provide an example to the public which will discourage others from similar unacceptable behavior. In a criminal action, the state is the party principally interested in seeing that justice is done, whereas in a civil action, the primary objective is to seek some remedy—such as payment of damages, granting a divorce, performance of a contract, or an injunction—against the defendant. In a criminal action, since the primary interests involved are those of the state, the prosecutor is given wide discretion in determining when a criminal charge will be sought and for what crime. In a civil action, as the primary interests involved are those of the plaintiff, the decisions about proceeding with the civil action, and for what remedy, are left to the discretion of the plaintiff.

A word of warning: never threaten criminal prosecution against someone who has caused injury to you if your motive is to get damages paid. Accepting money from someone who has committed a crime on the promise that there will be no complainant for the state in a prosecution is in itself an offense, known as "compounding a crime." In addition, for a public official to conceal a crime—or even to refuse to enforce a law that he is duty-bound to enforce—in exchange for some personal benefit also is a crime, known as "bribery."

In a case where there is no question of official duty involved, such as when a man has stolen a car from you, it is all right to promise not to sue him for damages in a civil action if he will return the car to you. Under these circumstances, you are not under any obligation to go to the police and file a complaint for the theft. In fact, you may accept payment for injuries and restitution of property without ever seeking legal redress. However, if someone else should file a criminal complaint against the man for having committed the theft, and you are later called upon to testify in court or answer police investigators' inquiries as to what took place, you must tell all you know about the incident. Although your private interest in obtaining return of your property has been satisfied, the public's interest in punishing criminal offenders has not. It must be recognized that the state's right to regulate conduct is independent of the individual's right to live without injury. The state, not the individual, must be satisfied in criminal matters.

Who May Sue in a Civil Action?

The right to sue is considered a fundamental right of all individuals and entities, and the suit can proceed provided there is some material, unresolved question of right or injury. The courts give significant deference to individuals to sue, for the courts do not want to forestall a person's opportunity to be heard and have his claims judged. This must be distinguished from the ability to obtain a favorable judgment of the court, which can be obtained only after a trial on the merits or an admission of guilt. It is always important to recognize the material distinction between a suit being filed and one upon which a judgment has been rendered.

Anyone may bring a lawsuit against a person who causes injury, whether the injury was intentionally or negligently inflicted. Specifically, for a civil action to be prosecuted, there need be only some existing entity which can be targeted by the processes of the law, and against whom the court's judgment can operate.[2] Examples of parties capable of bringing suit are individuals, corporations, governmental agencies, associations, partnerships, foreign governments, Indian tribes, aliens, convicts, and married persons acting as individuals to defend or prosecute their respective rights. With children, unborn infants, and people who are mentally incompetent, it is necessary for someone else to bring the suit for them; their guardians do this.

STATE OF CALIFORNIA
SUPERIOR COURT OF THE COUNTY
OF CONTRA COSTA

JOHN DOE)
Plaintiff)
) COMPLAINT FOR PERSONAL INJURY
vs.) AND PROPERTY DAMAGE
) CASE: 1234567
ANYTOWN FIRE DEPARTMENT)
Defendant)

Come now plaintiff John Doe, a citizen of the State of California, by way of complaint against defendant alleges as follows:

(a) That on May 28, 1987, defendant fire department, through its employee Fred Fireman, did operate a fire apparatus on or about 145 Irving Street, Anytown, which collided with plaintiff's parked vehicle, resulting in plaintiff's painful, disabling and permanent personal injury.

(b) That said apparatus operation was negligent....

James Q. Lawyer
Attorney for:
John Doe
(Address)
(Telephone Number)

A LAWSUIT CAN START with a civil complaint document, filed by the plaintiff, which states grievances against the defendant.

Who May Be Sued in a Civil Action?

In general, anyone who can sue can also be sued. Fathers can sue mothers, husbands can sue wives, brothers can sue sisters, even minor children may be

[2] Witkin, B.E., *California Procedure*, 3rd Edition, Bancroft-Whitney Co., 1985

sued. The simplified rule is that the court only needs to have an identifiable defendant upon whom its directives can be enforced to judge a suit.

There are many limitations on the right to sue, such as a time limit within which to bring the action. This time limit is commonly referred to as a "statute of limitation." Other limitations include the place in which the action may be brought, referred to as questions of "jurisdiction" and "venue." For suits against cities, most states and the Federal government, it also is necessary to comply with special statutory restrictions covering which courts can hear the suit, time limitations for filing actions, and special procedural requirements. Failure to comply with their myriad provisions will result in a court's dismissal of the action for lack of jurisdiction.

Telling a man that he cannot sue you is an incorrect statement. Although his cause of action may not be sustained by the court, it does not prevent him from filing the action against you. Only where the action is clearly without merit, on the face of the complaint, will a court refuse to consider the case. Note that if a person is the subject of a suit that is clearly without merit, or is undertaken purely for vindictive purposes, the person can sue the initiator for malicious prosecution and can recover defense costs and punitive damages. A general word of advice: "It pays to investigate before you litigate."

The prime requirement for a successful lawsuit is a good "cause of action." A cause of action arises when a legal right of the plaintiff is invaded by action of the defendant. Also, for a good cause of action, the defendant must lack defenses for his actions. Even though the plaintiff's rights have been invaded by the defendant, the plaintiff may not prevail for several reasons. The defendant, in a written response—called an "answer"—to the plaintiff's allegations may present several defenses to show he is not liable. The defendant may establish that the plaintiff never had a cause of action, or, if there had been one, there is none now. In addition, the defendant may indicate protection from liability under provisions of a "statutory immunity" law—a law preventing recovery of damages from an individual due to an overriding public interest or concern. A law protecting fire inspectors from liability for the negligent performance of an inspection would be an example of a statutory immunity.

In addition, a defendant may challenge the authority of the court to decide the case or enter an enforceable judgment. Such challenges to the authority of the court to render a decision include: 1. lack of venue—that is, action brought in the wrong court; 2. lack of notice of the existence and the nature of the proceedings; 3. existence of a concurrent or previously decided case covering the same allegations; 4. the court's lack of subject matter jurisdiction, meaning that the court is one of limited authority and cannot consider cases on this subject matter; and 5. the court lacks personal jurisdiction over the individual.

Another type of defense is one which admits the truth of the plaintiff's allegations, but introduces new matter to avoid the effect of the claims and to show that the plaintiff is not entitled to recover any damages. For example, the defendant may admit that he regularly drove his automobile over the plaintiff's land, but then may produce a written statement from the plaintiff proving he had the right to cross. Or, he may admit that his car collided with the back of the plaintiff's car, but then prove that he could not avoid the accident because the plaintiff backed into his lawfully parked vehicle.

Review Activities

1. You, as a minor child, have been injured in a traffic accident while riding in your parents' car. Whom could you sue to recover for your injuries?
2. What is the distinction between a civil action and a criminal action? Who can initiate each? Give examples.
3. Define the following terms:
   A. Lawsuit
   B. Plaintiff
   C. Defendant
   D. Jurisdiction
   E. Appeal
   F. Compounding a Crime
   G. Vindictive Lawsuit
4. Explain the meaning of "statutory immunity" as outlined in this chapter. Indicate why, as a matter of public policy, it would be important to grant to fire marshals statutory immunity for negligence in reviewing building plans for fire safety concerns.
5. Necessary to the success of a lawsuit is a good cause of action. In your own words, define "cause of action" and "good cause of action." Include appropriate examples of each.

Chapter Three

Criminal Actions

What Is "Criminal Law"?

"Criminal law" is that branch of jurisprudence which deals with the acceptable conduct required of all individuals and the penalties imposed for violations. In contrast to civil law, which focuses on the protection of individual rights and provision of redress for an individual's injury, criminal law focuses on the protection, safety, and welfare of the public, and on protection of all the people from the abuses of a few.

In addition, criminal law can be considered as the means whereby society punishes those who do not conform to its requirements. Punishment has multiple impacts on and benefits to society. As in *Commonwealth v. Ritter*, "There have been advanced four theories as the basis upon which society should act in imposing penalties upon those who violate its laws. These are: 1. to bring about the reformation of the evil-doer; 2. to effect retribution or revenge upon him; 3. to restrain him physically, so as to make it impossible for him to commit further crimes; and 4. to deter others from similarly violating the law."

Some of the important principles and individual protections provided by the system of criminal justice in the United States are:

1. Everyone is presumed to know the law, and thus is subject to punishment for its violation.
2. A person is presumed to be innocent until proven guilty, and is entitled to any benefit of the doubt where any reasonable doubt exists as to his guilt or innocence.
3. No person can be brought to trial until there has been a proper indictment, information, or complaint filed, with a copy of any and all charges presented to the accused.

4. An accused has the right to be represented by legal counsel at all critical steps of the judicial process. If an accused desires legal assistance but cannot afford to retain private counsel, public legal assistance, free of charge, must be provided.
5. The accused is entitled to a trial by a jury of disinterested peers, selected impartially from the general public.
6. The accused must be tried on the basis of the facts presented, and upon evidence legally collected, but not upon his character or prior conduct.
7. A person cannot be required to give testimony which could incriminate him. However, should someone choose to testify in his own defense, all questions posed must be answered truthfully.
8. No one can be tried for the commission of an act which was not a crime at the time the act was performed; nor can the punishment imposed be increased over that stipulated at the time of the act.
9. The accused cannot be put in jeopardy twice by the same jurisdiction for the same offense. Note, however, that this restriction does not prohibit imposition of punishment for individual steps of a crime which themselves are crimes, such as conspiracy, as well as for the completed crime.

What Is a "Crime"?

A "crime" is a wrong, directly or indirectly affecting the public, to which the government has annexed certain punishment, and which it prosecutes in its own name in what is called a "criminal proceeding." A crime also may be a violation of, or neglect to perform, a legal duty of such importance to the protection of society that the government takes notice of it and imposes a punishment.[1] Historically, the definition of what action is a crime has had its basis in the common law. Compilation of judicial decisions over centuries served to enumerate the actions constituting a crime and the proofs necessary for conviction. Today, with the on-going deliberations of the legislatures and the Congress, the body of common law crimes has largely been replaced with statutory descriptions. The United States Code and penal codes of the states now serve to name and describe nearly all crimes and their required punishments. These punishments can range from death or imprisonment to disqualification from holding an office of honor or public trust (for example, being an elected official, a judge, or even a fire fighter).

Crimes generally are classified into two categories: felonies and misdemeanors. The distinction between them is principally the type, character, and duration of punishment provided for their commission. Historically, felonies were considered serious enough to warrant forfeiture of life, limb, and/or lands and possessions to the felon's lord. Today, felonies are described as

---

[1] 22 *Corpus Juris Secundum* 2, West Pub. Co., 1961

crimes punishable either by death or by imprisonment for longer than one year. Misdemeanors consist of all other offenses except very minor infractions, called violations. The United States Code and the penal codes of the states typically state the classification of the crime.

Criminal justice is based upon a public desire to suppress people's unlawful intentions. For a crime to occur, there must be a union or joint operation of two critical elements: a prohibited act, and a specific intent to commit that prohibited act. Intent is shown by the circumstances surrounding the perpetration of the offense; that is, whether or not the prepetrator was of sound mind and enjoyed freedom of action. For example, an insane individual or someone under physical duress (such as when a man is pointing a gun at him, demanding action) would not have the requisite mental capacity or unbridled discretion to freely develop a criminal intent. Therefore, criminal acts occurring during such incapacity would not be considered crimes, as they were committed by individuals who lacked the critical element of free, voluntary, and unbridled intent to commit the prohibited act. Although this result might seem unsound, upon reflection it appears just, for criminal justice is premised upon punishing the unsocial actions and intentions of people rather than their unintended, negligent acts. Involuntary manslaughter is an example of criminal negligence for which a penalty may be imposed.

Minor exceptions are those criminal acts for which the perpetrator will be held strictly accountable. The very occurrence of the act is presumptive evidence of criminal liability. These crimes usually are limited to only minor infractions, are characterized as involving no moral turpitude or social stigma, and generally are punished by a fine. An example is the typical parking violation.

All persons are presumed to be of sound mind unless previously declared otherwise. The medical and legal ideas of insanity are essentially different. Insanity in medicine has to do with the departure of the indivudual from his natural mental state due to injury or disease. On the other hand, the focus for legal purposes is whether or not the individual had the mental capacity and moral freedom to do or abstain from doing the particular act. A person might be insane in the medical sense and still be held accountable for his criminal acts. For instance, an individual might be under the delusion that he is Moses reincarnate and yet understand perfectly the difference between right and wrong. If such a man were to deliberately murder someone under circumstances where his particular delusion had no bearing on the killing, he could be found sane insofar as his trial for murder was concerned.

Who Is Capable of Committing Crimes?

As a general rule, all persons are capable of committing crimes except those in the following categories:

1. Any child under the age of 14 for whom there is no clear proof that, at the time he committed the act charged against him, he knew the act was

wrong. Note that such a child could be tried in juvenile court as a delinquent, rather than as a criminal offender, and would be subject to punishment and rehabilitation under the juvenile justice system.
2. Idiots, lunatics, and insane persons.
3. Any person who commits the act or omission under a mistake of fact sufficient to disprove any criminal intent. For example, a person who injures someone in the honest and reasonable—although mistaken—belief that such action is necessary to preserve another's life is not guilty of any crime.
4. Any person who commits the act charged without being conscious of doing so. This might be a sleepwalker who performs a robbery.
5. Any person who commits the act or omission through misfortune or by accident, when it appears that there was no evil design, intention, or culpable negligence.
6. Except for a crime of homicide, any person under immediate threat of bodily harm. Such a person would not be guilty of any crime committed under the circumstances.

Who Is Liable to Punishment?

In general, the following persons would be held liable under the laws of a given state:

1. Any person who commits, in whole or in part, any crime within the state under whose laws he is being punished.
2. Anyone who commits robbery, larceny, or embezzlement entirely outside the state, but who brings the stolen property, or any part of it, into the state, or is found with it in the state.
3. Anyone who, although outside the state, causes, aids, advises, or encourages another person to commit a crime within the state.
4. Any person who commits perjury outside a state regarding a matter under consideration by a court or tribunal within the state.

Voluntary intoxication, whether by drugs, alcohol, or other substance, is no excuse for commission of a crime. An act committed by an intoxicated person is no less criminal by reason of his condition. However, whenever the existence of any particular purpose, knowledge, premeditation, deliberation, malice, motive, or intent is necessary to constitute a particular crime or degree of a crime, the jury can take intoxication into consideration in determining why a person committed the act, or whether he had the requisite intent to commit the act.

Parties to Crime

Parties to crimes are divided into two main categories: principals and accessories. A principal, within the historical meaning of criminal law, is any person present at the scene of a crime, and who directly commits the act constituting the crime or aids or abets its commission. Individuals who, by fraud, force, or contrivance, occasion the drunkenness of someone in order to cause him to commit a crime likewise are principals to the offense. Also included as principals to a crime are those who counsel, advise, or encourage minors, lunatics, or the insane to commit a crime.

An accessory is a person who in some manner aids in the perpetration of a crime. Historically, this designation of a participant in a crime had three classes: accessory before the fact (who, although not present at the time, encouraged the crime to be done); accessory at the fact (who was present at the time, but only as a lookout or assistant); and accessory after the fact (who, knowing that a crime had been committed, assisted the perpetrator to escape, or hid, comforted, or assisted him in any way to avoid arrest). Most states have merged classifications of accessories before and at the fact into the classification of criminal principal, so that a person in this category is considered an equal participant in the commission of the crime. Only accessories after the fact are considered separate offenders, not accountable for the underlying offense but only for aiding and abetting a known criminal offender.

Types of Crimes

Although crimes sometimes are classified according to the degree of punishment incurred as a result of their commission, they more generally are arranged according to the nature of the offense. The following, based upon analysis of *Black's Law Dictionary*,[2] is perhaps as complete a classification as is warranted here:

1. Common law crimes. Punishable under the common law, as distinguished from crimes established by statute.
2. Continuous crimes. Consisting of a continuous series of illegal acts. Examples are carrying concealed weapons or maintaining a public nuisance.
3. Crimes against the law of nations. Crimes which all nations agree should be punished, such as the crimes of murder and rape.
4. Crimes against property. Crimes of which the object is property. Examples are larceny and theft.
5. Crimes of omission. Failure to act when under an obligation to do so. An example is a driver's failure to aid a victim struck by his vehicle.
6. Crimes of violence. Involving the threat of, or actual occurrence of, bodily harm. Typical examples are murder, robbery, and assault.

[2] Black, H.C., *Black's Law Dictionary*, 5th Edition, West Pub. Co., 1979

7. Organized crimes. Perpetrated by organized groups, as contrasted with crimes planned and perpetrated by individuals. Common organized crimes are gambling, prostitution, and sale of narcotics.
8. Crimes against public property. Involving loss or damage to public property such as theft of public property.
9. Crimes against public justice. Injurious to the effective and orderly process of government. Examples are perjury, bribery, and contempt of court.
10. Crimes against the public peace. Destructive of the public order and an individual's peaceful enjoyment of personal and real property. Examples are riots, unlawful assembly, and libel.
11. Crimes against chastity. Involving deviant sexual behavior, such as adultery, bestiality, and incest.
12. Crimes against public policy. Against the general social interest of society, such as nuisances.
13. Crimes against the currency and against public and private securities. Acts which diminish public confidence in the authenticity and worth of valued documents. Examples are forgery and counterfeiting.

This list does not pretend to include every act which could be deemed criminal. However, it gives generally understood examples of the various types of activity that can constitute crime.

Review Activities

1. Write definitions for the terms "felony" and "misdemeanor." Distinguish between the types of punishment provided for each. Give an example of a type of crime, applicable to public health and safety, for each.
2. Define the term "crime," indicating its necessary elements.
3. In a written statement, explain the difference between an accessory and a principal as a party to a crime.
4. Describe in your own words society's reasons for criminal punishment. Do you believe such punishment is effective in view of current rates of crime?
5. Describe how the principles and protections of the criminal justice system are important to you as an individual and for society as a whole.

Chapter Four

The Judicial System

An Overview

In actuality, everyone is served by two judicial systems: the Federal system and the state or territorial system where he resides. Whether a case is brought in the Federal or state court is a matter of jurisdiction, depending on which court has the legal authority to render an enforceable decision over the matter.

Jurisdiction is of two kinds: over the person, known as "in personam," where the court has the authority to seize and incarcerate the individual; and over the thing, or "in rem," where the court's authority extends only to the property item and not the individual. A court must have at least one type of jurisdiction, but may have both, before it can begin to consider the legal and factual questions involved in a case.

The court's jurisdiction—its authority to hear a case—also is limited by the nature of the case, the court's status, and the monetary amount involved. For example, if a case involves a question of domestic relations, a specialized family law court has exclusive jurisdiction in the matter. A case involving small monetary amounts might be the exclusive original jurisdiction of a small claims court.

Jurisdiction also is classified as either original or appellate. The court in which a case must first be tried is one of original jurisdiction. All appeals of that original court's decisions are referred to the appropriate court having appellate jurisdiction. Because of the various jurisdictions of courts, a natural hierarchy exists among them. The court of original jurisdiction, typically called a trial court, hears the witnesses, reviews the written evidence, and decides the factual questions of a case. Based upon these adjudged facts, the court then applies the law to the facts and renders an appropriate judgment. Should an appeal be made, an appellate court would review the judgment of the original trial court to make sure that all court procedures were properly followed, that there was appropriate factual evidence to support the trial court's decision, and that the rules of law were properly applied. If these factors are present, the appellate court must sustain the judgment of the trial court.

Although there are two distinct court systems and types of law, Federal and state, trial courts of either system can consider cases involving the other's law. For example, a Federal trial court can adjudge a case involving questions of state law. A state court can consider cases involving Federal law. This is because the trial court's primary concern is the determination of fact, with subsequent application of the law to the adjudged facts. To assure proper application of the other's law, cases are often moved to the other court system for determination. In this way, the uniform application of the laws, whether Federal or state, is more likely to be assured.

State Courts

The existence and organization of a judicial system within a state is dictated by that state's constitution. The system typically is comprised of a supreme appellate court, district courts of appeal, trial courts (variously called "superior courts" or "supreme courts") and inferior specialized courts (municipal courts, justice courts, etc.). The scope of authority of the various courts is dictated by the state constitution as well as legislative enactments, and is periodically subject to change.

The Judicial Council

In a majority of the states there is a judicial council—an advisory body set up to study that state's judicial system, to compile statistics, and to do judicial research. Council deliberations frequently result in recommendations for improving judicial administration within the state.

The Supreme Court of the State

The supreme court is the highest appellate court of a state and the ultimate arbiter of that state's laws. It typically is composed of a chief justice and from four to six associate justices. The supreme court has presented to it all cases appealed from a state's courts of appeal or, in cases involving the interpretation of a state's law, from a Federal court. It customarily will review all cases involving the death penalty.

Decisions of the supreme court involving interpretations of state law must be followed by all other courts, Federal and state. The supreme court has the power to issue writs necessary to the exercise of its jurisdiction; such writs include *habeas corpus*, *mandamus*, *certiorari*, etc. (See the Glossary for definitions of these terms.)

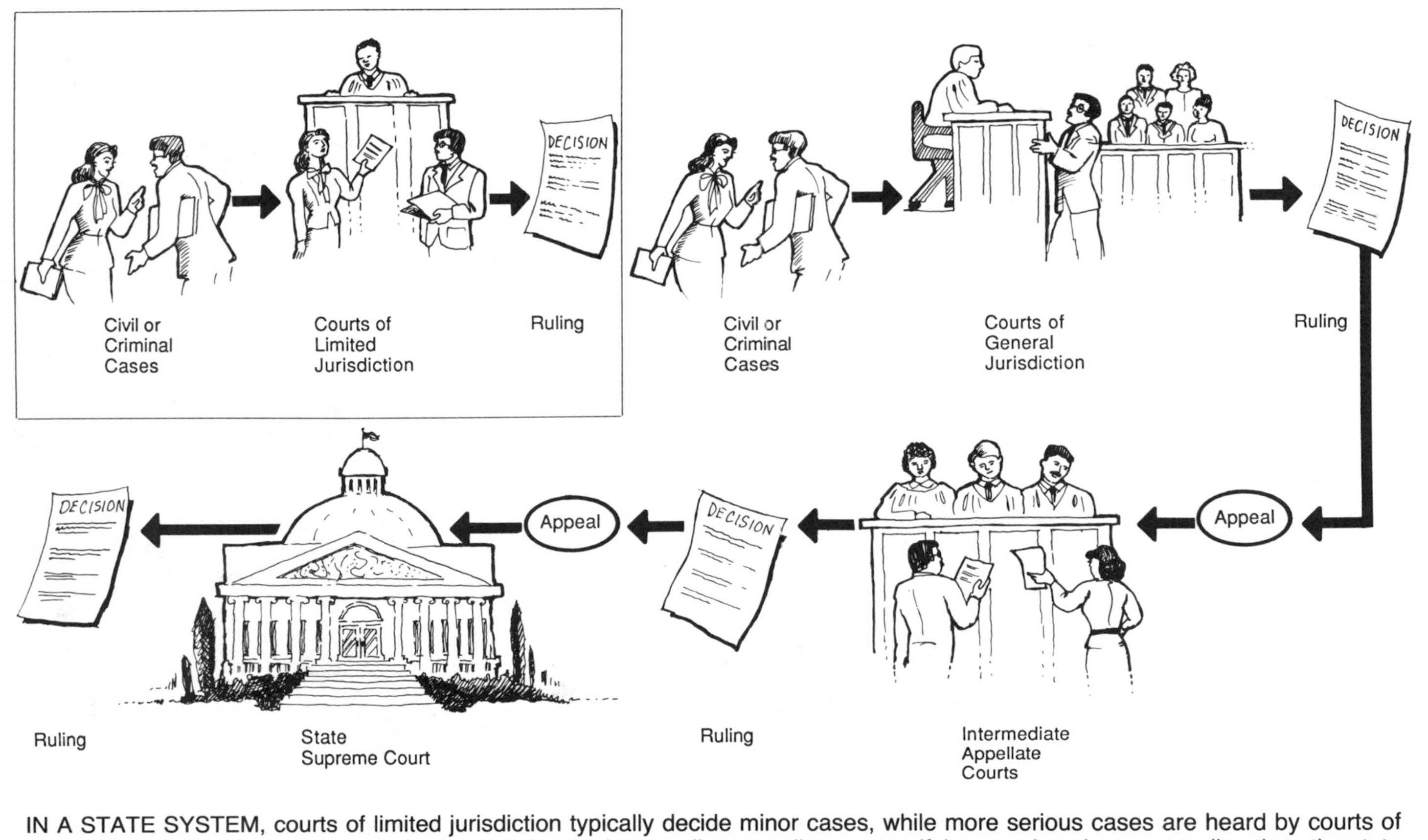

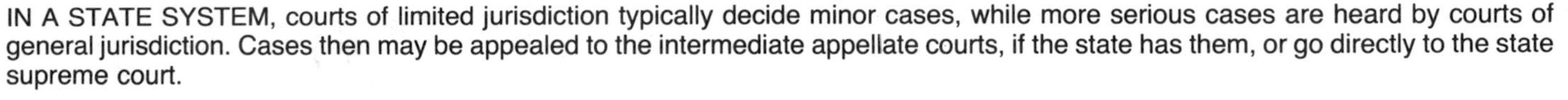
IN A STATE SYSTEM, courts of limited jurisdiction typically decide minor cases, while more serious cases are heard by courts of general jurisdiction. Cases then may be appealed to the intermediate appellate courts, if the state has them, or go directly to the state supreme court.

Courts of Appeal

A state can have several courts of appeal, each responsible for considering appeals of all cases emanating from a particular area of the state. These areas frequently are designated as judicial districts; each district covers a county or group of counties. These district courts consider all appeals of trial court judgments and orders unless a state supreme court assumes jurisdiction. In California, for example, there are six courts of appeal, each of which considers cases from one or several counties. The District I Court of Appeals, which sits in San Francisco, considers cases from several northern California counties.

Trial Courts

The trial courts of a state are its superior courts (entitled "supreme court" in New York). These courts are the principal judicial bodies for determination of civil or criminal liability. They are the most widely recognized courts in the state judicial system due to their use of citizen juries. In addition to consideration of general civil and criminal cases, trial courts have jurisdiction to consider cases involving individual and corporate taxes as well as probate, domestic relations, and juvenile affairs. They also have jurisdiction over prosecution for criminal actions involving felonies and certain misdemeanors.

Municipal Courts

Cities of large population generally have municipal courts to handle civil cases not involving large sums of money, and all misdemeanors except those committed by juveniles.

Inferior Courts

An inferior court is subordinate to a trial court and has special, limited jurisdiction to consider cases. Such courts include: 1. township justice courts; 2. city justice courts; 3. municipal courts such as "recorder's" and "police" courts in some towns; and 4. city courts. Inferior courts frequently are established to consider specialized problems; an example is a court for consideration of building and fire code violations, where specialized training of the court officials is required.

Reports of State Courts

A fundamental principle of the law is to stand by precedent. This means that when a court has laid down a principle of law as applicable to a certain set of facts, it will adhere to that principle. The court will apply the principle to all future cases in which the facts are substantially the same, regardless of whether or not the parties and property are the same.

To facilitate the implementation of this principle, decisions of a state's circuit courts of appeals and supreme court are compiled and published in volumes entitled *Reports*. Reference to these Reports guides trial and inferior courts in decision-making and, consequently, in the consistant, uniform application of the law.

In addition to the official publications of the various states, cases also are published by private companies in what is called the Reporter System. For example, decisions of the appellate courts of Iowa, Michigan, Minnesota, Nebraska, North and South Dakota, and Wisconsin are published in the *North Western Reporter*. Indexing and cross-referencing of cases on a particular subject facilitate comparison of the law among these states.

The Federal Judicial System

The existence and authority of the Federal judicial system is outlined by the United States Constitution. Article III, Section 1 of the Constitution provides: "The judicial power of the United States, shall be vested in one Supreme Court, and in such inferior courts as the Congress may from time to time ordain and establish." The Constitution also provides that the President shall appoint all judges of the Federal courts, who shall hold office for life, with removal only by impeachment for cause.

The United States Supreme Court

The highest court of the United States is the Supreme Court. Although it has original jurisdiction in some cases, such as those where one state sues another, most of its work involves the settling of appeals taken from the highest state courts (where a question of Federally protected right is at issue), and in the hearing of appeals from the circuit courts of appeals. The jurisdiction of the Supreme Court is limited, for it can consider only cases involving questions of international relations, Federal law, or Federally protected rights. Where none of these questions is at issue, the Supreme Court lacks authority to consider the case and must defer to the judgment of the appropriate state supreme court.

The Constitution of the United States does not expressly grant the Supreme Court of the United States the authority to declare invalid a law of Congress or of a state. However, in 1803 the U.S. Supreme Court asserted that it had this power, in the famous case of *Marbury v. Madison*. Chief Justice John Marshall based his reasoning on the premise that since the Constitution is the supreme law of the land, it is up to the highest court to hold invalid any attempt to circumvent the Constitution's provisions. To hold otherwise, Justice Marshall's opinion declared, would nullify the value of a written constitution and permit it to be altered by any statute which the current Congress or state legislature happened to pass.

Circuit Courts of Appeal

The United States is divided into 12 Federal judical circuits, in each of which is a circuit court of appeals. A circuit court considers all appeals of a district court's judgments within its district unless a case is one which can be appealed directly to the U.S. Supreme Court. A typical circuit encompasses from three to nine states; for example, the Seventh Circuit covers the states of Illinois, Indiana, and Wisconsin. Decisions of a circuit court of appeals establish legal precedent which all district courts within that circuit must follow. The law on a particular subject can vary among circuits, with resolution of the conflict achievable only through an appeal to and decision by the U.S. Supreme Court.

In addition to the territorial circuit courts, there is a specialized circuit court—the U.S. Court of Appeals for the Federal Circuit. Based in Washington, D.C., it has exclusive jurisdiction to consider appeals of decisions by the U.S. Court of Claims, the Court of International Trade, and the International Trade Commission.

The United States Tax Court

When an individual wishes to appeal a deficiency tax assessment (underpayment of taxes) made by the Commissioner of Internal Revenue, such an action is considered by the United States Tax Court. The Tax Court does not handle claims for over-payment of taxes; the latter are presented in either the U.S. Court of Claims or a U.S. district court.

Court Procedure

The procedural steps through which a court considers a case depend upon whether the action is of a criminal or civil nature. The general elements of the respective procedures are as follows.

Criminal Procedure

1. If a citizen comes to a city attorney, county counsel, or district attorney to protest the acts of a neighbor, the attorney can have a complaint drawn up and served on the potential defendant, with a copy filed with the court. Besides being served with a copy of the complaint, the defendant would be notified of the date to appear in court. When the violation is of a serious nature requiring immediate action, and when probable cause exists, the attorney could petition the court for issuance of a "bench warrant" directing the immediate arrest of the defendant.
2. For certain crimes, an immediate arrest may be made by a peace officer or citizen, with or without a warrant. Crimes permitting immediate arrest typically are limited to felonies or serious misdemeanors. Following arrest, the accused is "booked," an administrative process

in the police station whereby the name of the accused, crime charged, and other pertinent facts are entered into the police department's criminal record or "blotter." The arrested person's fingerprints and photograph also might be taken at this time.

3. Next, the arresting party or police officer will present the facts of the alleged crime to the city attorney or public prosecutor. If the prosecutor believes a crime has been committed and that prosecution of the arrested party is warranted, a document called an "information" will be issued. This document charges the arrested party with the crime.

   Alternatively, the attorney or public prosecutor could present the facts to a grand jury which, should the evidence warrant it, would issue an "indictment."

4. At the "noticed" hearing[1] or immediately following arrest, the accused is brought before a court for arraignment. At arraignment, the accused is publicly advised of the charges against him and of his constitutionally protected rights, and is requested to enter a "guilty" or "not guilty" plea. If the accused pleads guilty, he is sentenced. If he pleads innocent, he is retained in custody or released on bail pending the next court session, called a preliminary hearing.
5. At the preliminary hearing, the city attorney or public prosecutor presents the general elements of the case against the accused. The court then determines if there is sufficient evidence to warrant conducting a formal trial. Note that this proceeding is not to determine guilt or innocence, but is a preliminary determination by the court whether or not continued prosecution of the accused is warranted. If the evidence is sufficient, the accused is "bound over" for trial.
6. At a trial requiring a jury, the jury will be selected by a process called *voir dire*, unless the accused waives his right to trial by jury. This entails the prosecuting and defense attorney each questioning each prospective juror on his background and impartiality in the subject matter involved in the trial. When the number of jurors and alternates required to constitute a jury have been accepted by both attorneys, the jury is formally impaneled to hear the case.
7. The trial proceeds as follows:
   A. The prosecuting and defense attorney each can make an opening statement which outlines his perspective of the case and the evidence he will present.
   B. Prosecutor presents his witnesses and evidence to prove his case.
   C. Defense attorney cross-examines the prosecutor's witnesses. Prosecutor may then elect to reexamine his witnesses.

---

[1] In a "noticed" hearing, a person is formally advised or warned of forthcoming proceedings in which his interests are involved.

D. Defense attorney presents his witnesses and evidence.
E. Prosecutor cross-examines the defense attorney's witnesses. Defense attorney may then elect to reexamine his witnesses.
F. Defense attorney and prosecutor then may each make a closing statement.

8. The judge charges the jury to consider the evidence presented, to determine the ultimate facts of the case, and to decide on the guilt or innocence of the accused. The judge also will instruct jurors on the law which should be applied in the case.
9. After considering all the evidence and applying the instructions on the law issued by the judge, the jury will determine a verdict. If the verdict of "guilty" is found, the judge will impose the sentence. If the required number of jurers cannot agree upon a verdict, the defendant must be acquitted of the charges.
10. The accused may appeal the verdict or ask for a new trial on various grounds which tend to show that he did not have a fair trial. The state, however, cannot appeal if the accused is found "not quilty."

Civil Procedure

The principal steps in a civil suit are:

1. The plaintiff files a complaint with the appropriate court, outlining his grievances and requesting that the court impose an appropriate remedy against the defendant.
2. A summons is issued by the court, demanding that the defendant respond within a specified time period to the allegations contained in the complaint. The response can be in the form of an "answer," which might deny any or all portions of the complaint, or a "demurrer," in which the defendant alleges that he is not, for legal reasons, obliged to "answer" the complaint.
3. "Discovery" is conducted. In this process, the plaintiff's and defendant's attorneys exchange evidence, and take depositions under oath of opposing parties and experts in an effort to narrow the issues in question.
4. The court conducts preliminary, pre-trial examinations and settlement conferences to encourage the parties to settle their dispute prior to trial.
5. A trial is conducted, either before a judge and jury or before a judge alone.
   A. Plaintiff's attorney opens the case by calling witnesses and presenting evidence to prove his contentions.
   B. Defendant's attorney cross-examines plaintiff's witnesses in an attempt to discredit their testimony or bring out facts damaging to plaintiff's case.
   C. Defendant's attorney then calls his witnesses to testify, and they in turn are cross-examined by the plaintiff's attorney.

D. Summation is conducted, each attorney giving closing statements indicating why the judge or jury should rule in his client's favor.

6. The judge instructs the jury in matters of law. The jury then is charged to determine the facts of the case, and to apply the law to the facts in their determination of a verdict. At least three-fourths of the jury must agree in the judgment.
7. If the trial was conducted without a jury, the judge will determine the facts of the case as well as the ultimate judgment.
8. There may be a request for a new trial because of some serious error committed during the legal process; or an appeal may be made to a higher court because of some errors in law or judging the evidence.
9. If the plaintiff is successful, he will attempt to have the judgment enforced. This might be an order of the court for the defendant to perform or refrain from performing some act, or might be an award of money damages. The order also might require levying against the defendant's property, whereby the property can be seized by the sheriff and sold to satisfy the amount awarded the plaintiff.

Review Activities

1. In outline form present, in your own words, the principal steps in a civil suit.
2. Describe the procedure for the prosecution of a criminal action by creating an imaginary situation and, on a step-by-step basis, applying to it the principal steps involved.
3. Under what circumstances must the United States Supreme Court defer to the judgment of a state supreme court?
4. Under what circumstances can an appellate court overrule a trial court? Under what circumstances can it not?
5. What checks are provided in the criminal justice process to assure that the constitutionally protected rights of the accused are not violated? Describe two of them.
6. Describe the function of the following steps in the criminal justice process:
   A. Booking
   B. Arraignment
   C. Preliminary Hearing
   D. *Voir Dire*
   E. Jury Instructions

Chapter Five

Organization of Fire Departments

Background

While the first organized efforts to extinguish fire antedate by centuries the modern fire department, it must be remembered that it has been only a relatively short time—approximately 125 years—since fire fighting was taken over by public authorities. In the vast majority of instances, fire suppression services were performed by private, volunteer organizations. However, over time this method of providing public fire suppression services proved ineffective for larger communities. "Notwithstanding the efficiency of private forms of fire protection, city governments justified their entry into fire fighting in the mid-19th century as a response to market failure in the private sector; [their] inability to control violence by competing fire fighters . . . . The supposed market failure terminated a period in which fire fighting services were provided solely by private, volunteer effort. For the first 200 years following the settlement of North America, private mutual societies and then private clubs produced fire protection services. But within some 20 years (1853-71), the claim that volunteers' lawlessness required public enterprise carried the day. Most large cities legislated the volunteers out of existence and replaced them with tax-paid municipal departments . . . ."[1]

The advent of the modern municipal fire department, with its vast resources in the technology of fire prevention, detection, extinguishment, and control, has given rise to myriad legal problems. At the outset, it is well to have a clear idea of the considerations involved in the creation and organization of fire departments.

---

[1] McChesney, Fred S. "Government Prohibitions on Volunteer Fire Fighting in Nineteenth-Century America: A Property Rights Perspective", *Journal of Legal Studies*, Vol. XV, Jan. 86, pp. 69-92

The creation and composition of a fire department—whether it is wholly or partially under the control of a state, or under the exclusive control of a local governing body—is a matter generally determined by local law. This law is principally enumerated in state constitutions and statutes, and in municipal charters and ordinances. For example, a municipal fire department is an agency of local government, maintained for the benefit of the community; its composition, powers, and duties are defined by city charters and ordinances. Other departments, for example rural fire protection districts, are organized under state laws and are independent, autonomous entities having their own policy-making and administrative authority. Thus, it is important to recognize the source of powers of a fire department or district, the limits of its authority, and the requirements for its formation and operation.

Because volunteer fire departments comprise the vast majority of organized fire suppression entities,[2] their organization and operations must be understood.

Volunteer Fire Departments

Organization

About nine out of ten fire fighters in the United States are volunteers in rural and smaller community fire departments. Although many of these departments have been established under state laws authorizing the formation of fire districts, some have been organized as individual fire companies within departments of incorporated communities. For example, Sections 14825-27 of the California Health and Safety Code provide that fire companies in unincorporated towns may be organized by entering a certificate with the county recorder. Signed by the company's foreman or presiding officer and by the secretary, the certificate shows the date of organization, the name of the company, names of the officers, and the roll of active and honorary members.

It is important that fire departments formed under a type of law similar to California's comply with the requirements to the letter. This is in order to qualify their apparatus as "authorized emergency vehicles" and to provide their members with the protections of common law and statutory immunities. The mere installation of a red light and siren on a fire truck does not entitle its driver to exceed the speed limits, go through stop signs, or otherwise disregard traffic laws; such exemptions are accorded only to operators of "authorized emergency vehicles." When a fire company has not been organized as prescribed by law, or if its legal status has not been properly main-

---

[2] NFPA has identified more than 26,000 volunteer and mostly-volunteer fire departments within the United States. The total number of departments is estimated at 28,260.

Career and Volunteer Fire Fighters in the U.S. by Population Protected (1985)

	Career		Volunteer		Total	
Population Protected	Number	Fire Fighters per 1,000 People	Number	Fire Fighters per 1,000 People	Number	Fire Fighters per 1,000 People
1,000,000 or more	29,800	1.36	150	.01	29,950	1.37
500,000 to 999,999	23,100	1.36	11,200*	.66	34,300	2.02
250,000 to 499,999	24,400	1.55	2,400	.15	26,800	1.70
100,000 to 249,999	36,500	2.32	1,950	.12	38,450	2.44
50,000 to 99,999	31,950	1.30	9,000	.37	40,950	1.67
25,000 to 49,999	34,050	1.15	22,450	.76	56,500	1.91
10,000 to 24,999	34,850	.94	66,900	1.81	101,750	2.75
5,000 to 9,999	16,250	.61	109,650	4.14	125,900	4.75
2,500 to 4,999	3,850	.18	172,000	8.19	175,850	8.37
under 2,500	3,750	.17	443,750	19.57	447,500	19.74
TOTAL	238,500	1.00**	839,450	3.52**	1,077,950	4.52**

Source: NFPA Survey of Fire Departments for U.S. Fire Experience, 1985
*Reflects data from several county fire departments which include considerable numbers of volunteer fire fighters.
**Result weighted by community size.

tained, its members would not qualify for the immunities that protect them from personal liability. Members also might be deprived of the benefits which workers' compensation laws ordinarily provide for line-of-duty injuries.

Existance of such a situation is not purely hypothetical. At a meeting of the California State Board of Fire Services in July 1975, it was reported that at least 32 "organized fire companies" and 81 volunteer fire departments throughout the state had a questionable legal status for performance of fire services because of possible noncompliance with statutory provisions for organization. Every fire department should check to make sure that it has complied with the laws for its formation and continuation. Such compliance is necessary to give its members all possible protections, immunities, and benefits provided under the laws of the jurisdiction.

Exemptions, Benefits, Powers, Duties

In general, volunteer fire fighters have the same rights, duties, powers, and privileges as paid fire fighters. These rights apply provided the volunteers' names have been duly enrolled on the fire department register and recorded as required by law. Also, individuals must be in full compliance with all requirements for retention of member status. Once these prerequisites have been met, volunteers have the rights to enter upon property to gain access to fire, to make rescues of endangered persons, and in general to carry out the duties imposed by their department rules.

Under most circumstances, volunteers whose names have been properly enrolled can qualify for workers' compensation benefits, as can their dependents, in the event of injuries or death incurred while acting within the scope of their employment.[3] However, the fact that a person is not registered with a volunteer company will not preclude him from obtaining workers' compensation benefits in the event of injury: many states have adopted laws which provide that anyone who volunteers to help suppress a fire, when requested to do so by an officer or employee of a public agency, is deemed an employee of that agency and is entitled to compensation.

Volunteers must be provided with all the safeguards required for paid fire fighters by state and Federal laws. These include protective clothing, self-contained breathing apparatus, and other safety features, as well as effective fire fighter training. The provider of these safeguards is determined by local law.

---

[3] For example, California Labor Code, Section 3361, 1984, provides workers' compensation benefits to volunteer fire fighters under the following statutory provision: "Each member registered as an active fire fighting member of any regularly organized volunteer fire department, having official recognition, and full or partial support of the government of the county, city, town or district in which the volunteer fire department is located, is an employee of that county, city, town or district. . .and is entitled to receive compensation from the county, city, town or district . . . ."

Benefits available may include: permitting volunteer fire fighters to become special fire police with full power to regulate traffic, control crowds, and exercise all other police powers necessary to facilitate the fire company's work at a fire or any other emergency; exempting their associations from apparatus title and registration fees; and making fire companies eligible for low interest state loans in order to purchase equipment. Other statutes may prohibit an employer from terminating a volunteer for missing work while responding to a fire call; fire fighter relief associations may be entitled to a tax on fire insurance premiums paid within the district; and fire fighting apparatus and equipment may be exempted from property taxes. (*Harmony Volunteer Fire Department v. Pennsylvania Human Relations Commission*)

A Volunteer's Liability

Just because a volunteer serves without pay does not make him immune from liability for negligence or willful misconduct. Even though he has volunteered his services, responded to a fire call in his private vehicle, and purchased his protective clothing with his own funds, he is not protected from liability for his careless actions. The volunteer is afforded the powers and many of the benefits of the paid professional, but such grants come with commensurate responsibilities and potentials for personal liability. Unless the volunteer is afforded statutory or common law immunity from liability, should the volunteer fail to exercise due care he may be held accountable for the consequences.

For example, some states have made special provisions whereby a volunteer fire fighter may install a warning light, siren, etc. on his own car, and also have granted him certain privileges under emergency response conditions. Lacking such provisions, he must obey all the rules of the road that apply to ordinary civilian drivers. Obviously, no fire fighter under the influence of any intoxicant should operate any emergency vehicle nor engage in any fire suppression or rescue operations. His condition not only could imperil his own life and that of his co-workers, but also could make him personally liable for any harm that results.

The issue of liability for fire suppression operations is discussed in Chapter 6, for fire prevention activities in Chapter 8, and for rescue or first aid service in Chapter 13. When reviewing the material, it is important to remember that the standard of care of the volunteer and paid fire fighter is the same. The court does not distinguish a person's duty based upon his level of compensation, but only upon the nature of his position.

Liability Insurance

Volunteer (and other) fire fighters have been known to serve alcoholic beverages at some of their get-togethers. For this reason, it is advisable for volunteer associations to obtain liability insurance. This would cover any liability that might arise from a motor vehicle accident involving negligence of an intoxicated member who had just left a social event sponsored by the association. Where alcoholic beverages have been served on fire department property, whether or not in violation of department rules but with the acquiescence of town officials or district directors, it is highly likely that the officials or directors would become defendants in any subsequent litigation.

Although the fire company or department receives financial support from the community and is partially regulated by the community's elected officials, this does not mean it is automatically covered by the community's liability insurance. Any time coverage is in question, confirmation from the community's risk manager or carrier is advised.[4] (*Cook v. City of Geneva*)

Fire Districts

In addition to the fire protection afforded by volunteer fire companies within municipalities, discussed previously, as well as by municipal, county, and state fire departments, fire services can be furnished by organized districts. Water, fire, and special service districts exist under the laws of many states. There are reported to be more than 3,000 such districts functioning within the United States.[5]

Organization of Fire Districts

"A fire district is a territorial subdivision of the state bounded and organized under the authority of the legislature for the government purpose of providing protection against fire within its limits . . . . Although composed of a part of one or many towns it is in substance a quasi-municipal corporation of definitely restricted [purpose and] powers." (*President of Williams College v. Williamstown*) A fire district has been variously described as a political subdivision of a county or state, a public corporation, or even a municipal corporation. The designation is important only in determining the interrelationship among the district and other public entities and the status of the entity under state law.

---

[4] An independent, incorporated, volunteer fire company, which was a defendant in a personal injury action, requested court declaration that the company was covered under the city's liability policy. Although the company was virtually free to carry on its internal affairs without interference from the city, when the alarm sounded in this case, the city's fire chief assumed command over volunteers' operations. Upon the chief's assumption of command, the company effectively became an element of the city and thus was protected under its liability policy, the court ruled.

[5] Antieau, Chester J., *Independent Local Government Entities*, Local Government Law Series, Matthew Bender Pub. Co., 1983, Vol. 3A, p. 30D-3

A fire district must be formed in strict accordance with applicable state statutes, which require close scrutiny.[6] Typical provisions of such authorizing statutes might include necessary procedures for formation, elections, and boards of directors; area limitations; financial and taxing authority; contractual authority; changes in boundaries; and means for annexation, reorganization, and dissolution. Procedures for district formation must be complied with in detail. An example of the importance of following the statutory prescriptions exactly is an Illinois case where a fire protection district was held to be "fatally defective." The reason was that, when formed, the district had included an island in the Mississippi River which actually was located in the State of Missouri. (*People ex rel. Curtain v. Heizer*)

A city, or part of it, might be included in a fire protection district, although some statutes forbid this unless the city is annexed as part of the district. In Illinois, where a city and a fire district had overlapping boundaries, the city was ordered to cease performing fire protection services, to stop levying taxes for this purpose, and to let the fire district perform the services in the area. (*People v. Edge O' Town Motel*) In some states, fire districts are established as corporations; they are considered public corporations in New York (*Seyer v. Schoen*) but not in Washington, while in Rhode Island they are held to be quasi-corporations.

---

[6] For example, under California law "any one of the following alternative methods may be used to form a district":

> (a) The supervising authority [such as the county board of supervisors]... may determine, by resolution, that certain territory not served by a regularly organized fire department of a city and not served by an existing fire department of a district is in need of fire protection and should be formed into a district. The board of supervisors shall then proceed to hold a public hearing and an election [to determine whether or not the residents want a fire district].
> (b) A petition for the formation of a district which is signed by registered voters in the proposed district equal in number, at least, to 25 percent of all the votes cast within the proposed district for all the candidates for governor at the last preceding general election at which a governor was elected. The board of supervisors shall then proceed to hold a public hearing and an election [to determine whether or not the residents want a fire district].
> (c) If a petition for the formation of a district presented to the board of supervisors is signed by registered voters in the proposed district equal in number, at least, to 51 percent of all the votes cast at the last preceding general election at which a governor was elected, the board of supervisors, after holding a public hearing... may, by resolution, declare a district formed without recourse to an election.
> (d) If a petition for the formation of a district presented to the board of supervisors is signed by owners of real property in the proposed district equal to at least 51 percent of the assessed valuation of real property within the proposed district as ownership and valuation appear on the last equalization assessment roll, the board of supervisors after holding a public hearing... may, by resolution, declare the district formed without recourse to an election. California Health and Safety Code Section 13825, 1961

Some provisions of the fire district organizational laws relate to the terms of district officers, their compensation, and what constitutes a quorum; the methods by which the board can take action (ordinance, resolution, or motion); and how it can meet its financial needs. The latter provisions include the procedure for obtaining additional revenue to meet expenses. More revenue might be needed due to increased costs mandated by state regulations, or higher premiums charged for workers' compensation insurance.

Where prescribed by statute, the procedures for dissolution of a fire district must be complied with strictly. Although board members are authorized to operate a district, they do not have an incidental, implied, or inherent power to dissolve a district. Again, "municipal corporations [fire districts] only possess expressly granted powers, and powers implied from or incident to those expressly granted powers. Any reasonable doubt concerning whether or not a municipal corporation [fire district] possesses a given power must be resolved in the negative. Further, where the legislature has authorized a municipal corporation [fire district] to exercise a power, and prescribed the manner in which it should be exercised, any other manner of exercising the power is denied to it." In this case, a Missouri fire protection district was held not dissolved by the unilateral action of its board of directors. Dissolution could be achieved only through compliance with the statutorially prescribed dissolution procedures. (*State ex rel. Crites v. West*)

District Board Representation

Where members of the board are determined by election, each board member should represent an equal number of registered voters within the district—the "one man, one vote" concept. Alternatively, where board members are elected to represent a segment, each individual segment of the district should consist of an approximately equal number of voters. However, where board members are appointed to their position, there is no obligation for each to represent an equal number of voters. For example, in *Clark County v. City of Las Vegas*, a state statute provided for formation of a metropolitan fire department consisting of the City of Las Vegas and a large segment of Clark County. However, the statute providing for the metropolitan district designated the city's commission (city council) to manage, administer, supervise, and control the resultant department. In holding the statute unconstitutional, the court noted that the entire governing board of the resultant district and a majority of its budget committee would thus be elected only by residents of the city. As a result, there would be significant disenfranchisement of county residents. Such restricted representation would be in violation of the equal protection clause of the United States Constitution.

Powers of the Board of Directors

All the authority given to a fire district can be exercised only by its board of directors, constituted as a board; an individual board member has no power whatsoever. Without authorization by his board in a public meeting, he cannot, for example, give orders to the fire chief, or purchase an item for the fire department and then seek reimbursement for the amount he has expended.

Fire protection districts traditionally possess only those powers expressly granted them by statute, those reasonably implied to effectuate the basic powers, and those incidental to the accomplishment of the purposes for which the districts were formed.[7] As stated in *Glenview Rural District v. Raymond*, "Fire protection districts, like all municipal corporations, derive their existence and all their powers from the legislature. They possess no inherent powers and must be able to point out the statute which authorizes their acts."[8]

However, where provided, such powers can be very broad. They can include, for example, the right of eminent domain; the right to sue and be sued; to acquire real property; to employ legal counsel; to establish a water system for fire protection needs; to set qualification standards for fire fighters; and to operate not only fire equipment but also ambulances, rescue, and support vehicles. Powers also may include the right to enter into automatic or mutual aid response contracts, establish a civil service system, and provide members of the district fire service with insurance (health, life, and disability) coverage.

A fire district can be empowered to maintain membership in local, state, and national fire protection organizations on behalf of its fire department, and to pay reasonable expenses. These include authorized travel for any of its members or employees to attend professional or vocational meetings. The district also may be empowered to clear land of brush, grass, rubbish, or other combustibles. This can be achieved by order to the owner, by contract with private cleaners, or by the district's own personnel. Reimbursement of any reasonable expenses involved is the responsibility of the land owner.

Rules

Even though the volunteers in a fire district may have their own association and bylaws, these bylaws are not the rules of the district's board of directors. Typically, the board is the head of the fire department and the fire chief is its administrator. The board is responsible for setting the policy of the district; the fire chief, as administrator, is responsible for implementing the board's policy decisions.

While volunteers may set their own standards for membership in their association, again these are not binding upon the district's board in making

---

[7] Antieau, *op. cit.*, p. 30D-9

[8] Action of a fire district in promulgating a fire protection code including automatic sprinkler system requirements was held *ultra vires*, as an activity not authorized by the legislature.

appointments. The board may (and should) confer with the officers or committee of the volunteers' association before appointing the fire chief, making promotions, filling vacancies, setting salary schedules, terminating members, and setting policies on leaves, vacations, type and frequency of drills, uniforms, apparatus, etc. Nevertheless, the board has the sole authority for making the final decisions in these matters.

Records

Good records are essential to avoid legal pitfalls that might arise at later dates. For example, medical records should be in the personnel file of each member, starting with the entrance physical examination and including all annual checkups required thereafter. Each student fire fighter also should be given a medical exam to determine that he has not suffered a prior injury that might later be confused with a compensable one sustained at a fire. Without good medical records, there is no way of knowing whether the cause of a disability preexisted or not.

Records for fire apparatus and equipment maintenance also are important. Failure to make regular checks of the air or oxygen supply in breathing apparatus and resuscitators might be relevant in case of a claim against the district because of an empty cylinder or contaminated air. Was each empty cylinder replaced following a fire or rescue run? Driver certification records for operators of emergency vehicles, as well as certificates of emergency medical (paramedic) training and hazardous material training, could be useful in defense against lawsuits alleging negligence by the board for failing to assure proper training and credentialing of members.

Public Meetings

State laws regulate the degree to which public business must be conducted openly. Some laws contain exceptions to the full open meeting concept, for example when personnel affairs such as appointments, promotions, dismissals, or other matters of discipline are being discussed. Consultation with legal counsel over issues involved in litigation also may take place during closed meetings. Laws may forbid holding meetings in any facility that prohibits access to women and minorities, such as a private men's club. Legal provisions also may stipulate that a member of the public attending a meeting cannot be required to register his name, fill out a questionnaire, or do anything else as a condition of attendance.

District board members should not conduct business informally, such as over coffee at a local cafe. Gatherings of board members, especially of a majority, can present the image of a secret meeting in violation of open meeting laws. Rather, members should confine discussion of board business to direct conversations between individual members, or group discussions in the forum of a board meeting of which notice has been posted.

Executive Sessions

If an executive session is held to discuss confidential matters, personnel for example, the meeting room should be cleared of visitors or the board should retire to a separate room. Only the fire chief, employee representative, or legal counsel may be present in case he is needed to confer on the matter under discussion. Attendance by other individuals outside the board should be strictly prohibited. The board member who holds the title of secretary can take any notes deemed desirable, although these need not become part of the official minutes which must be made accessible to the public. Only minutes of the public session must be kept and, upon reasonable notice, must be made available to any member of the general public for review and copying. Minutes of the public meeting should show the time the board went into executive session, the reason for the executive session, and the time the public session was reconvened, and should contain a brief note on results of the executive session.

Legal Counsel

A fire district needs good legal counsel to protect its interests in the event of litigation. It may be the county's counsel or district attorney who provides such service. If there is any question about the availability of public counsel, it might be best for the fire district to retain private counsel to handle district matters. Since it is not uncommon for a county to bill a district for the services of its legal staff, retention of private counsel might bring more immediate attention to the needs of the district, at little additonal cost.

Should litigation occur, counsel should be supplied with all records, minutes, names of witnesses, photos, and all data potentially relevant to the suit. Release of such information also may be required for retaining insurance coverage should a claim be made against the insurance carrier. For example, if the suit is based upon some action that the board was alleged to have taken, such as voting on an issue in executive session, minutes of that meeting should be produced to show that the board reconvened after its executive session and voted on the matter in open public session. If no such minutes exist, or no notation was made as to sessions beginning and ending, the actions of the board could be construed as illegal. Detailed records could eliminate this potential problem.

A suit against a fire district did occur, with the conditions outlined above present. The plaintiff alleged that he was granted a two-year contract to supervise the fire chief. The transcript of a tape recording of the executive session showed that board members did not assent to such a contract, despite the fact that the minutes failed to show the board reconvened in general session or voted on the contract. Two of the four board members present at the meeting later signed affidavits that the board never did reconvene in general session or vote on the matter. On the basis of these facts, the trial court ruled that the fire district had to pay the plaintiff two years' salary for work he never performed. The court reasoned that parole (word of mouth) evidence from a board member was admissible to dispute the minutes and other evidence. Appropriately detailed minutes would have precluded this outcome.

Directors' Liability

Directors are liable the same as private individuals for their own negligent or wrongful acts. They also are liable for the wrongful expenditure of public funds. While they cannot make gifts of public funds or property, the law generally permits gratuitous transfer of property between governmental units. Directors cannot make private use of public property, such as borrowing the fire department's pickup truck for transport of personal property.

Although the board may use district funds to print a fact sheet on a bond issue, it is illegal to use public funds for publicizing board candidates or financing political activity. Directors, as public officers, are not personally liable for the acts of members of the fire department unless, as pointed out in the discussion on liability of superior officers, the directors instigate or direct the acts, participate in them, ratify them, or assent to them.

Both criminal and civil liability can result from such actions as accepting bribes, favors, gratuities, or "kickbacks" on the purchase of fire equipment, or selling property to the district at an inflated value. Whether such transactions took place directly or through "dummy" parties, the director involved would be liable to the district for the full loss and any costs incurred in obtaining recovery.

Conflict of Interest

Many states have laws that deal with conflict of personal or financial interest of public officials. In the case of *Terry v. Bender*, the court decided that the mayor had a financial interest in the continuation of a contract between the city and a special attorney it employed. This situation tainted the contract with illegality and rendered any payments to the attorney illegal, as being contrary to law and the public interest. The court stated: "A public office is a public trust created in the interest and for the benefit of the people. Public officers are obligated. . .to discharge their responsibilities with integrity and fidelity. Since the officers of a governmental body are trustees of the public wealth, they may not exploit or prostitute their official position for their private benefits. When public officials are influenced in the performance of their public duties by base and improper considerations of personal advantage, they violate their oath of office and vitiate the trust reposed in them, and the public is injured by being deprived of their loyal and honest services.

"It is therefore the general policy of this state that public officers shall not have a personal interest in any contract made in their official capacity . . . . A transaction in which the prohibited interest of a public officer appears is held void both as repugnant to the public policy expressed in the statutes and because the interest of the officer interferes with the unfettered discharge of his duty to the public. The public officer's interest need not be a direct one, since the purpose of the statutes is also to remove all indirect influence of an interested officer as well as discourage deliberate dishonesty. Statutes prohibiting such 'conflict of interest' by a public officer should be strictly enforced."

It is difficult for a person to be a fire fighter, who must take orders from the fire chief during drills and at fires, and concurrently be a member of the board of directors which holds the fire chief accountable for the efficiency of drills and for command at fire scenes. To avoid potential criticism and conflict, it would be advisable for a director either to resign from the department or to take a leave of absence from the ranks upon being elected to the board of directors.

Dual membership, within the same or different departments, also can be prohibited by statute. In *State ex inf. Gavin v. Gill*, a newly elected fire district board member challenged the state statute prohibiting dual membership[9] as violating his First Amendment rights. The Missouri Supreme Court sustained the prohibition, stating: "The First Amendment protects the right to participate in political activity and to seek and hold public office, but political involvements of public employees may be restricted if a sufficient public interest is shown .... The Legislature well might conclude that an employee of one governmental unit should not be eligible to serve as a member of the governing board of another. Given this conclusion, there is no less restrictive alternative. In the appellant's situation the possibility of conflict is clear. He might be reluctant to approve as director a contract or agreement which would affect his duties as fire fighter. If it need be shown that the public interest is compelling, we conclude that the test has been met."

Other Considerations

In addition to the applicable procedural pointers presented in the final chapter of this book, a number of suggestions pertain especially to directors of fire districts. These pointers might assist in avoiding or solving some legal problems, including:

---

[9] "No person holding any lucrative office or employment under this state, or any political subdivision thereof. . .shall hold the office of fire protection district director .... When any fire protection district director accepts any office or employment under this state or any political subdivision thereof, his office shall thereby be vacated and he shall thereafter perform no duty and receive no salary or expenses as fire protection district director." Missouri Revised Statutes, Section 321.015, 1978

1. Public Information.

Conflicting information to the news media and the public must be avoided. Policy should be to refer inquiries to the fire chief in regard to fire department activities and to the board chairman on matters relating to policy decisions. It should not be necessary for an officer of the department to call a board member to determine whether or not he should use department apparatus to remove a cat from a tree, replace a flag pole halyard, pump out a flooded cellar, and other non-fire protection requests. Such policies should have been established by the board, recorded, and made available to the public and media upon request.

2. Promotional Examinations.

Although this subject is discussed in more detail in a later section, a word of caution is in order at this point. It is best for the board to arrange to have promotional examinations developed and administered by professionally qualified persons who are not associated with the fire department; to do otherwise could open the door to charges of partiality at the very least. The extent of the board's participation should be to interview the candidates who pass the examination for the position of fire chief, and, other factors being equal, select the individual who heads the list. Promotions in the lower ranks should follow a comparable procedure with the chief selecting the individual(s) who head the eligibility list.

3. Discipline.

The board's rules for the department should provide that the fire chief be given the discretion to suspend any member for serious violations of the rules, and to impose lesser penalties for minor infractions. Provisions also should be made for a grievance procedure and hearing, by right, at the request of the disciplined member. Such procedures should be fully detailed, adopted by the board, and published. A written "findings of fact" justifying the discipline imposed should be developed and kept for future reference. This document can be used to rebut any allegations of discrimination concerning activities punished and discipline imposed.

Fire-Police Combination Departments

In a number of communities the functions of both fire and police protection are being performed by the same public employees. Arrangements include requiring fire fighters and police to perform the others' duties; or consolidating the traditional fire and police department funtions and assigning them to a public safety department. In the absence of a statute prohibiting such arrangements, there should be no legal objection to such undertakings as they usually are considered matters for local determination. Under the applicable laws of Illinois, for example, the City of Peoria had the power to adopt an ordinance combining to some extent the duties of the police and fire departments and to require appointive officers or employees to perform duties in both departments. (*Hunt v. City of Peoria*) Authority to order consolidation may be vested by statute in a specific municipal official.[10] In *O'Leary v. Town Manager of Arlington*, the authority of the town manager to order fire-police department consolidation was sustained as an enumerated power of that official under that form of local government.

Consolidation can present unusual questions of labor-management relations, especially where the fire and police departments are combined into a public safety unit. For example, in *Vair v. City of Ravenna* a challenge was made against the authority of the city to require fire fighters to operate the police radio. The Ohio Supreme Court held that the requirement violated a state statute which said fire departments "shall protect lives and property of the people in case of fire. . .and such other duties as are provided by ordinance." The court held that the city could not require fire fighters to operate the radio, as the statute intended to limit "such other duties" to be limited to those incident to an individual's position, such as fire fighting.

A 1975 decision by the Supreme Court of Vermont, *IAFF Local 2287 v. City of Montpelier*, held that the city could not compel fire fighters and police to be represented by the same union. This was the court's finding even though the city claimed that it would save time and would be more expedient to negotiate both contracts at one time, with one bargaining representative. Again, in a Michigan case, *Kalamazoo Police v. City of Kalamazoo*, that state's Fire Department Hours of Labor Act was held inapplicable to former fire fighters who had transferred to public safety officer status. "The hours of work provision here at issue does not violate the. . .statute because the hours of labor act (24-hour shifts) only applies to traditional fire departments, not to the hours of work of public safety officers in fully integrated public safety departments." The work hours negotiated in a collective bargaining agreement were found applicable.

---

[10] Massachusetts General Laws Annotated provides: "The town manager. . .may reorganize, consolidate or abolish departments, commissions, boards or offices under his direction and supervision, in whole or in part, [and] may establish such new departments, commissions, boards and offices as he deems necessary and make appointments to such boards, subject to the approval of the board of selectmen. He may in connection with such transfer of such powers or duties transfer the duties and powers of one department, commission, board or office to another and may, with the approval and consent of the finance committee, transfer the appropriation of such one department, commission, board or office to another." Massachusetts General Laws Annotated, c.503 Section 15(b) of Acts of 1952

Organization of Municipal Fire Departments

Obligation to Provide Fire Protection Services

Without a state constitutional or statutory requirement,[11] state or local government has no obligation to provide public fire protection services. Rather, the provision of such services is considered to be permissive. "The legislative body of a city *may* provide fire engines and all other necessary or proper apparatus for the prevention and extinguishment of fires."[12] "The board of fire commissioners, with the approval of the board of trustees, *may* organize and maintain fire, hose, protective, and hook and ladder companies, whenever in its judgment the public interests require."[13]

Failure to provide fire protection services does not constitute a deprivation of anyone's civil rights. "The Constitution creates no positive entitlement to fire protection." Rather, it is a "charter of negative liberties; it tells the [government] to let people alone; it does not require the Federal government or the state to provide services, even so elementary a service as maintaining law and order [or providing fire protection]." (*Jackson v. Byrne*)

A statutory obligation to provide fire protection services need not mean the organization of a fire department. A statute may permit the community to provide such services through contract with a volunteer fire company or may simply mandate that the community pay for fire suppression services rendered by any fire department. For example, in *Rockwood Volunteer Fire Department v. Town of Kossuth*, a volunteer fire department sued the town for the costs of providing fire suppression services to extinguish three local fires. In sustaining the fire department's right of recovery under state law,[14] the court noted: "In enacting . . . , the Legislature exercised its power to determine what public policy shall be with respect to fire protection, and required such town boards to provide such protection in one of three ways: They could establish their own fire department, contract with another for such service, or pay for the services of any fire department responding to calls within their boundaries."

---

[11] "The legislative body of a city organized under general law shall establish a fire department for the city. The fire department shall be under the charge of a chief who shall have had previous training and experience as a fireman. The other members of the fire department shall consist of paid firemen or such companies of volunteer firemen as the legislative body may determine." California Government Code Section 38611 (1955)

[12] California Government Code Section 38600 (1949)

[13] New York Village Law Section 10-1004

[14] "Any town failing to provide for. . .a fire department and fire fighting apparatus and equipment for extinguishing fires in such towns shall be liable for the services of any fire department in fighting fire and appearing to fight fire in such town upon request." Wisconsin Statutes Section 60.29/18m

The statutory obligation to provide fire protection also may arise in response to public petition. For example, "Under Indiana law, if a majority of real property owners who live within an area of a township which lies outside a municipality petition the township trustee and legislative body to provide fire protection services in that part of the township, then the petition shall be granted and the trustee and the legislative body must proceed without delay to provide fire protection." (*Ayres v. Indian Heights Volunteer Fire Department*)

In the absence of a contractual or statutory obligation to provide fire suppression and rescue services, is there a private obligation to pay for them? Is there an obligation to reimburse the community for its costs in providing those services? *District of Columbia v. Air Florida* states: "The general common-law rule in force in [most] jursidictions provides that, absent authorizing legislation, the cost of public services for protection from fire or safety hazards is to be borne by the public as a whole, not assessed against the tortfeasor whose negligence creates the need for the service .... Where emergency services are provided by the government and the costs are spread by taxes, the tortfeasor does not anticipate a demand for reimbursement."[15]

Where a fair and sensible system for spreading the costs of an accident already is in place, such as via taxes, the courts generally will not sustain cost recovery. (*City of Flagstaff v. Atchison*)[16] However, the courts appear open to sustain cost recovery from negligent parties where statutes or local regulations provide for such recovery, as in *D.C. v. Air Florida* and *People v. Wilson*.

Organizing the Fire Department

Remember that, unless directed by statute or a city charter provision, the organization of a municipal fire department is exclusively a local discretionary matter.[17] In addition, unless a statute or charter provision directs the performance of a specific fire department service, determination of the na-

---

[15] The District of Columbia filed action to recover costs of emergency services and cleanup required as the result of an airplane crash.

[16] The city attempted to recover expenditures arising from the evacuation of people due to derailment of railroad tank cars transporting liquefied petroleum gas. "Here governmental entities themselves currently bear the cost in question, and they have taken no action to shift it elsewhere. If the government has chosen to bear the cost for reasons of economic efficiency, or even as a subsidy to the citizens and their business, the decision implicates fiscal policy; the legislature and its public deliberative processes, rather than the court, is the appropriate forum to address such fiscal concerns."

[17] "The legislative body of a city organized under general law shall establish a fire department for the city." California Government Code Section 38611

ture and extent of fire protection services is reserved to local government.[18] (*Save Our Fire Department v. City of Appleton*)

Within the governmental entity, the organizational structure of fire protection services is exclusively a local decision. Typical ways services are organized are the Fire Chief and Command Structure; the Fire Administrator and Staff Structure; and the Public Safety Organization. The only requirement is that the organizational structure be in accordance with Federal and state law, city charter provisions, and local ordinances.[19]

Are Fire Fighters "Public Officers" or "Employees"?

While there is no legal distinction between a "member" and an "employee" of a department, a controversy exists as to whether a fire fighter is a "public officer" or simply an "employee." Fire fighters often are vested with discretionary powers regarding the nature and extent of inspections, the enforcement of laws relating to fire prevention, and duties pertaining to fire suppression and investigation. Therefore fire fighters may be considered officers. In other jurisdictions, where fire fighters are subject to close supervision by their superiors and are vested with only limited discretionary authority, they usually will be considered public employees.

"The following principal guidelines may. . .be used in making this determination: (1) The position was created by law and involves continuing and not occasional duties. (2) The holder performs an important public duty. (3) The position calls for the exercise of some portion of the sovereign power of the State. (4) The position has a definite term for which a commission is issued and a bond and an oath are required." (*James v. Prince George's County*) Applying these four points in the *James* case, the court ruled that operators of fire apparatus and ambulances were *not* considered public officials covered by official immunity while operating emergency vehicles.

To determine the question of status, in each instance reference must be made to the charter or statute governing the service, as well as to pertinent judicial decisions. In paid fire departments, while there are some judicial

---

[18] A citizen initiative to amend the local charter and mandate retention of paramedic service was upheld for election by voters. A charter provision by initiative can overrule a city council decision.

[19] "Local laws embodied in statutes, charters and ordinances control as to the creation and composition of the fire department, its functions and powers, the appointment and advancement in grade of members, the regulation of the hours of service of the members thereof, fiscal matters relating to the department, and management generally." McQuillan, Eugene, *The Law of Municipal Corporations*, Callaghan and Co., 1980, Section 45.03, p. 18

decisions, charter provisions and statutes to the contrary, many states hold that fire fighters—even fire chiefs—are not public officers. Why is the determination important? The designation "public officer" or "employee" will establish the power and authority of the fire fighter; the procedure required for his removal from office, for abolition of his position, and for reduction or assignment of his salary; some workers' compensation rights; and the potential liability of the city and individual for negligent acts.[20]

Are Fire Fighters "Peace Officers"?

Because fire departments often are responsible for detecting fire code violations and for apprehending arsonists, the question arises "Are fire fighters peace officers?" Do theyhave the authority to carry concealed weapons by virtue of their position? Are they required to complete training courses in the handling of firearms and to learn the application of the laws of search and seizure, arrest, and interrogation in those states where all "peace officers" must take such training?

The determination of whether or not a fire fighter is a "peace officer" is covered by state statute. For example, under California Penal Code Sections 830.3 and 830.31, deputy state fire marshals and fire investigators in organized fire departments, charged with enforcement of the state's arson or insurance fraud laws, are "peace officers" empowered to carry concealed weapons, make criminal investigations, etc. All other members of the state fire marshal's office and fire departments are not considered peace officers.

The designation "fire police" is not the same as "peace officer." In many states, fire fighters designated fire police are granted power to regulate traffic, control crowds, and otherwise facilitate the fire department's work at a fire or other emergency.[21]

However, a peace officer is empowered to detain or arrest perpetrators of crime, enforce search and arrest warrants, serve supoenas, and enforce the state's penal code with all the authority of a police officer. A fire policeman generally lacks such authority.

---

[20] "Before a governmental representative in this state is relieved of liability for his negligent acts, it must be determined that the following independent factors simultaneously exist: (1) the individual actor, whose alleged negligent conduct is at issue, is a public official rather than a mere government employee or agent; and (2) his tortious conduct occurred while he was performing discretionary, as opposed to ministerial, acts in furtherance of his official duties." (*James v. Prince George's County*)

See also "Damage Suits Against Public Officers," *University of Pennsylvania Law Review*, Vol. 129, pp. 1110-1188.

[21] See example, 35 Pennsylvania Statutes Section 1201.

In any event, all fire fighters whose duties include the enforcement of codes and ordinances, irrespective of their qualification as peace officers, should become familiar with the proper procedure for making searches, seizing evidence, etc. Furthermore, no member should carry a concealed weapon until he is positive that he is in compliance with the weapons laws of his jurisdiction and has had all the required training.

Federal Impact on Fire Departments

Insofar as management controls are concerned, there was a time when a fire chief had to consider only his city's charter, ordinances, and state statutory provisions in directing fire department operations. Today, however, the interests of the Federal government can be seen in most aspects of the fire service, from employee selection to fire fighter personal safety, and from affirmative action to disability pensions.

Acronyms for some of the Federal acts and agencies which affect the fire service include FEMA (Federal Emergency Management Agency), DOT (Department of Transportation), FLSA (Fair Labor Standards Act), NBS (National Bureau of Standards), HHS (Department of Health and Human Services), OSHA (Occupational Safety and Health Administration), ATF (Department of Alcohol, Tobacco and Firearms), NTSB (National Transportation Safety Board), and HUD (Department of Housing and Urban Development). Along with the Department of Justice, the National Fire Academy within FEMA and the Wage and Hour Division of the Department of Labor, in enforcing the FLSA, probably have the most direct impact on fire department operations.

In addition, the United States Constitution and the various Federal laws prohibiting discrimination in employment have had major impact. These protections are enforced principally by the Equal Employment Opportunity Commission (EEOC), the Department of Justice, and the Federal courts. Their respective decrees and judgments have impinged upon fire department operations in many respects which will be discussed later.

Fundamental Constitutional Considerations

"There is no constitutional right to be a fireman. When one voluntarily elects to be a fireman he swears, voluntarily, to obey the orders and directions of his superiors to the best of his ability. The oath does not contain the reservation that obedience is conditioned upon personal agreement with such order. Necessarily he waives certain rights when he takes the oath." (*Krolick v. Lowery*)

Throughout the sections that follow, various Federally protected rights will be discussed along with ways the courts have examined these protections in the fire service setting. As no right is absolute—not even a Constitutionally protected one—it is the balancing of the individual's and government's interests that defines the limits of a fire department employee's protected rights. Particular attention is directed to how the courts have struck a balance between the interests of the individual in exercising his protected rights, and those of the employing agency in discharging its duties.

Wages and Working Hours

The Fair Labor Standards Act of 1938 was amended in 1974 to include public employees engaged in fire suppression activities. The changes became effective January 1, 1975. However, in a 1976 United States Supreme Court decision, *National League of Cities v. Usery*, the amended provisions were held unenforceable on the local government level. The court ruled that the Commerce Clause of the United States Constitution does not empower Congress to enforce the minimum wage and overtime provisions of the Fair Labor Standards Act (FLSA) against the states and their subdivisions. In a 1985 decision, *Garcia v. San Antonio Metropolitan Transit Authority*, the Supreme Court reversed itself and held that Congress *could* regulate the minimum wage and overtime provisions for state and local government employees. The provisions were applicable retrospectively, the extent depending upon the nature of service involved.

What constitutes the "working time" for which the FLSA is applicable? "In *Jewell Ridge Coal Corp. v. Local No. 6167*, the [United States] Supreme Court had to determine what constituted working time which made up the work week within the meaning of the Fair Labor Standards Act.... The Supreme Court set forth a three-step test [for] determining the essential elements of work. The court stated that the time must involve: (1) a physical or mental exertion whether burdensome or not; (2) the exertion is controlled or required by the employer; and (3) the exertion is pursued necessarily and primarily for the benefit of the employer and his business." These tests have been applied to various aspects of public service operations to determine whether they constitute work and thus are subject to minimum rates of compensation and overtime pay.[22] (*Ebright v. City of Whitehall*)

"While the Fair Labor Standards Act does set basic minimum wage and overtime pay standards, and regulates the employment of minors, there are a number of employment practices which the Act does not regulate. For example, the Act does not require: vacation, holiday, severance or sick pay; meal or rest periods, holidays off, or vacations; premium pay for weekend or holiday work, pay raises or fringe benefits; a discharge notice, reason for discharge, or immediate payment of final wages; any limit on the number of hours of work for persons 16 years of age and over. These and similar matters

---

[22] Time for roll call and meals constituted compensable time within the meaning of the FLSA.

are determined by [state law or] by agreement between the employer and the employees or their authorized representatives."[23]

Applying the provisions of the FLSA will be the subject of uncertainty and possible change for some time to come. For example, there are proposals to amend the Act as it applies to special detail work within the fire service.[24] Therefore, periodic follow-up on the implementing provisions of the FLSA, as published by the Wage and Hour Division, Department of Labor, is advised.

Safe Working Conditions

The Occupational Safety and Health Act (OSHA) was signed into law on December 29, 1970 and became effective on April 28, 1971. Employers, as defined and covered by the Act, are required to provide safe and healthy working conditions as outlined in the various OSHA implementing standards.[25] As originally promulgated, municipal fire departments were not subject to OSHA regulations for subdivisions of states; that is, fire departments of counties, cities, townships, etc. were not considered "employers" as defined by the OSHA statute. However, a fire department, if part of a political subdivision, now is required to comply with OSHA standards where its state has assumed responsibility for administration and enforcement of OSHA regulations within the state borders.[26] Such inclusion was a condition of the Federal government for compensating individual states for enforcement efforts. Many other states encourage voluntary compliance with the standards but have not granted any state agency the authority to enforce the regulations against local governments.[27]

Even if not a political subdivision—not organized under some statutory authority—a fire department is covered under OSHA if there is an employer-employee relationship. This would be the case for private, for-profit fire departments or volunteer departments. Even though no one in a volunteer fire department or on its board of directors receives pay or fees of any kind, the fire station still might be considered a "place of employment" and subject to the state's enforcement of the OSHA regulations. Because many fire departments have been inspected for compliance with OSHA standards and have had penalties assessed for violations, it is prudent to obtain copies of the OSHA standards and comply with them.

---

[23] "State and Local Government Employees Under the Fair Labor Standards Act," U.S. Department of Labor, WH Publication 1459, May 1985

[24] Conference Report, Report 99-357, U.S. House of Representatives, Fair Labor Standards Amendments of 1985, Nov. 1, 1985

[25] "Each employer...shall furnish to each of his employees employment and a place of employment which are free from recognized hazards that are causing or are likely to cause death or serious physical harm to his employees." 29 United States Code Section 654(a)(1)

[26] See example in Maryland Annotated Code, Article 89, Section 29(b), 1973.

[27] See North Carolina General Statutes 595-148.

An interesting right protected by the Occupational Safety and Health Act is the right of an employee to choose not to perform an assigned task because of a reasonable apprehension that death or a serious injury might result. Specifically, this right is applicable where the worker believes there is no less drastic alternative than refusing to work. This right has been sustained by the courts. They note that the law does not require employers to pay workers who refuse to perform assigned tasks in the face of imminent danger, but simply provides that in such instances the employer may not "discriminate" against any employee who reasonably refuses to work.[28] (*Whirlpool Corp. v. Marshall*)

Employment Discrimination

The Civil Rights Act of 1964, as amended by the Equal Employment Opportunity Act of 1972 (Title VII) and enforced by the Equal Employment Opportunity Commission (EEOC), has been instrumental in bringing about changes in the customary criteria for selection of recruits for the fire service. These laws prohibit discrimination in employemnt based upon race, color, sex, religion, or ethnic origin. Title VII covers all state political subdivisions which have 15 or more employees, but does not apply to elected public officials, their personal assistants, or policy-making appointees.

See "Supreme Court Upholds Workers' Right to Refuse Dangerous Assignment under OSHA Regulations," *Fire and Police Personnel Reporter*, April 1980, Issue 64, p. 5, for further details.

The following showings are necessary to establish a case of discrimination:

1. Plaintiff is a member of a suspect or protected class of individuals (such as a minority member of a sexual, racial, religious, or ethnic origin class);
2. Plaintiff applied and was qualified for a job for which the employer was seeking applicants;
3. Despite plaintiff's qualifications, the applicant was rejected;
4. After rejecting plaintiff, the job remained open and the employer continued to seek applicants with plaintiff's basic qualifications. (*McDonnell Douglas Corp. v. Green*)

---

[28] The court noted: "Such circumstances will probably not often occur, but such a situation may arise when (1) the employee is ordered by his employer to work under conditions that the employee reasonably believes pose an imminent risk of death or serious bodily injury, and (2) the employee has reason to believe that there is not sufficient time or opportunity either to seek effective redress from his employer or to apprise OSHA of the danger." (*Whirlpool Corp. v. Marshall*)

It is not necessary for the individual to actually submit an application. It need only be shown that such an effort would be useless inasmuch as the defendant had indicated that any application submitted by the plaintiff would be rejected outright.[29] (*Franklin County Sheriff's Office v. Sellers*)

Entrance Qualification Testing

The test for determining a violation of Title VII involving applicant qualification testing is twofold. First, the selection procedure must be shown to "adversely impact" members of a race, color, sex, religion, or ethnic group. "A selection rate for any race, sex or ethnic group which is less than four-fifths [or 80 percent] of the rate for the group with the highest rate will generally be regarded by the Federal enforcement agencies as evidence of adverse impact." (*Berkman v. City of New York*) Second, where adverse impact is claimed, the defendant (the employing agency) must show that the test was job-related—that is, that the test was "valid"[30] because it accurately selected those applicants who could effectively perform the job. Once an applicant demonstrates that the qualification testing procedures were discriminatory, that he is a member of the suspect or protected class, and that following rejection the employer continued to seek applicants for the position, the case for illegal discrimination has been established.

For example, a written entrance examination in which 52 percent of the white fire fighter applicants pass versus only 29 percent of the black or minority applicants would be considered to have adversely impacted the minority members. In other words, 80 percent of 52 percent equals 41.6 percent which materially exceeds the minority passage rate of 29 percent. Thus adverse impact has been shown. A physical agility test (such as a hose drag or ladder climb) wherein 46 percent of the male versus 0 percent of the female applicants pass would be considered to adversely impact female applicants.[31] (*Berkman v. City of New York*) When an adverse impact is shown, the defendant employing agency must demonstrate that the examination and each of its elements bears a valid correlation between success on the individual

---

[29] Female applicant for sheriff's position was told that only men were being considered for the position.

[30] Unless a logical relationship can be shown between the test material and what a person will do on the job for which he is being tested, the test lacks content validity. Validation can consist of requiring fire fighters to take a battery of tests and then compare their scores with the department's evaluation of the work. Another method is to put the test data away for a year, and at a later date compare it with the job performance ratings.

[31] The physical agility test for New York City fire fighter applicants was held to discriminate against women. Because the abilities that the tests attempted to measure were not the most observable abilities of significance for successful performance of the job, the test lacked subject matter validity.

exam question or physical test and performance of the fire fighter job.[32] Should the employer fail in his obligation and should the plaintiff be a member of the protected class, the court has the authority to direct an affirmative remedy—that is, hiring of the individual.

Tests for knowledge or physical abilities must cover basic elements for successful completion of training and job performance; the tests must not be based on attributes which could be achieved upon successful completion of basic training.[33] (*McCrea v. Cunningham*) The tests must show that the individual could, or could not, successfully complete basic training and the probationary period. Promotional examinations also must test for elements that are essential for successful higher level performance. Where the qualifications for successful performance of a fire captain's position were "assessing the scene of a fire and issuing the appropriate oral orders," paper and pencil examinations that essentially tested for written communication skills were held invalid. (*Firefighters Institute for Racial Equality v. City of St. Louis*) The Uniform Guidelines on Employee Selection Procedures (1978), 29 Code of Federal Regulations, Chapter XIV, Part 1607, outlines the EEOC's recommended procedures for test criteria selection, test formulation, and validation. Reference to provisions of these guidelines is highly recommended.

Having a job-related test is not enough to satisfy Title VII; the test also must be properly administered. In *Albermarle Paper Co. v. Moody*, the United States Supreme Court found that the tests in question were improperly administered because they were given selectively; they were not validated on younger, inexperienced, or minority applicants, but on experienced white workers. Moreover, the job performance ratings given by the supervisors to

---

[32] Entrance examinations can serve two purposes: to determine if an applicant possesses the necessary minimum qualifications for successful job performance; and to determine the best-qualified of all those possessing the necessary minimum qualifications. Both purposes also can be discriminatory. However, an examination to determine the best-qualified will be the subject of greater judicial scrutiny. "The (EEOC) Guidelines provide that rank-ordering should be used only if it can be shown that a higher score...is likely to result in better job performance.... If test scores do not vary directly with job performance, ranking the candidates on the basis of their scores will not select better employees.... The frequency with which such one-point differentials are used for important decisions in our society, both in academic assessment and civil service employment, should not obscure their equally frequent lack of demonstrated significance." (*Berkman v. City of New York*)

[33] The Supreme Court of Nebraska upholds promulgation of minimum visual acuity standards for fire fighters that refer, in part, to NFPA 1001. Despite availability of equipment to correct eyesight, a department could reasonably conclude that potential loss of corrective lenses during fire fighting could present an unreasonable safety risk to other fire fighters, thus permitting reasonable uncorrected visual acuity requirements.

validate the exams were not deemed objective enough to identify the specific criteria used in making the ratings.

Entry Qualifications

In addition to entrance examinations, the Equal Employment Opportunity Commission has ruled that other qualities of applicants cannot be considered because of their unvalidated relationship to successful job performance and their adverse impact on minorities. The EEOC has ruled that a requirement that applicants who have served in the armed forces must have an honorable discharge is not a valid prerequisite. The reason is that twice as many blacks receive dishonorable discharges as whites, indicating "racism" as the most significant factor in this disparity. The commission also has ruled that arrest records cannot be used to disqualify applicants, as experience showns blacks are arrested substantially more frequently than whites in proportion to their numbers. Also, unless it can be shown that facial hair needs to be regulated as a safety measure, it is discriminatory to bar applicants who have beards, mustaches, goatees, etc., since these are much more prevalent among minorities then among whites; they sometimes are also symbols of racial pride or cultural heritage. A 5-foot, 7-inch minimum height requirement was held violative of Title VII (*Davis v. County of Los Angeles*) because it disproportionately disqualified Mexican-Americans,[34] who tend to be shorter than the minimum.

Where discrimination based upon entry qualifications has been proven, the remedies available to the court can be significant to include the imposition of minority quotas. For example, in the case of *Carter v. Gallagher*, the court ordered the civil service commission to give absolute preference to 20 minority applicants on the next examination for fire fighter. Further, the court ordered the commission to lower the minimum age from 20 to 18 years; raise the maximum age from 30 to 35 years (at least until the 20 minority applicants were hired); remove the requirement of a high school diploma or an equivalency certificate; delete any reference to arrest records; accept an applicant with a felony or misdemeanor record if the felony is more than five years old or the misdemeanor more than two years old; and, finally, disregard entirely the time interval between the conviction and application for employment if the crimes cannot be proven to have a bearing on the applicant's suitability as a fire fighter.

Background Investigations

Regulations restricting the dissemination of information on a person's criminal history have been issued by the United States Department of Justice

---

[34] "Title VII forbids the use of height requirements which have discriminatory effect unless the employer meets the burden of showing that the requirement has a manifest relation to the employment in question."

and by state agencies. Although criminal justice agencies can be given such information for legitimate purposes, including preemployment checks, it is unlawful to pass the information along to anyone else except under specific conditions. Therefore it would be futile to request that the police department run an FBI check on an applicant's fingerprints if it is illegal for the police chief to tell the fire chief what the check discloses.

However, the regulations typically permit the disclosure, under certain conditions, of the arrest of an employed fire fighter. The regulations also provide that other public agencies may be given the records of arrests (up to one year back) and records of convictions. As indicated earlier, under equal employment opportunity guidelines, the use of criminal records as grounds for disqualifying applicants is not looked upon with favor. However, should a background check disclose that an applicant is wanted for parole violation, or is an escaped convict or otherwise a fugitive from justice, the fire department would not be required to accept that person's application.

State Legislative Regulation of Local Fire Departments

Just how far a state legislature may go in regulating the membership, hours, wages, and management of local fire departments without infringing upon the rights of "home rule" for municipal government is a question that has received diverse answers. In West Virginia, a statute placed paid municipal fire departments under civil service and authorized the appointment of members of the civil service commission by chambers of commerce and local boards of trade. When the mayor of Charleston, West Virginia refused to appoint a commissioner, he was compelled to do so by a writ of mandamus, despite his contention that the statute was unconstitutional. The court held void only that portion of the law which limited the selection of fire fighters to the "two great political parties" and upheld the remainder of the law.

Working Hours

As noted previously, a 1985 United States Supreme Court decision now makes the minimum wage and overtime provisions of the Fair Labor Standards Act applicable to public fire departments. Note that the FLSA only establishes minimum requirements and does not preclude individual states from mandating more favorable working hours or compensation. In addition, the Act does not direct the type of shift arrangement or schedule (two, three, or four platoons, or three 8-hour shifts, or 9- and 15-hour shifts) that must be used by a department; rather, individual states and communities are free to select the shift arrangement desired provided its terms comply with the FLSA. Such shift arrangements must comply with the state constitution and other state law, and must not be unconstitutional exercises of authority reserved to a particular level of government. (*City of Alexandria v. Alexandria Fire Fighters*)

For example, a Michigan law giving fire fighters one day off every four days, and 20 days' vacation, was held unconstitutional as an unwarranted interference with local self- government and an invalid exercise of the police power. The court said "It is well settled that a city's fire department is distinctly a matter which concerns the inhabitants of the city as an organized community apart from the people of the state at large, peculiarly within the field of municipal activity and local self-government." However, in Louisiana, in *New Orleans Fire Fighters v. New Orleans*, the court held that the state legislature may require that vacations be granted with pay.

Residency Requirements

To some, it appears desirable for fire fighters to live in the communities where they work. This is not merely to enable them to vote on bond issues, pension proposals, and other matters affecting their welfare, but also to make them readily available in the event of a major emergency. Moreover, any increase in taxes that would inevitably follow their obtaining higher salaries, shorter working hours, better pensions, more fire trucks, new fire stations, etc., would be borne by the fire fighters as well as by other resident taxpayers.

Several state supreme courts and the United States Supreme Court have ruled on the validity of residency requirements. In *McCarthy v. Philadelphia Civil Service Commission*, the U.S. Supreme Court ruled that Philadelphia's residency requirement was valid and not an infringement upon a person's constitutional right to travel interstate. The decision meant that a fire fighter with 16 years of service did not have a protected right in his position; movement of his residence from Philadelphia to New Jersey, in violation of the city's residency requirement, provided a valid basis for his discharge.

Also, a residency requirement has been held not in violation of a person's constitutional right of familial association. A restriction on the location of a residence does not amount to a constitutional infringement, for it does not regulate family affairs directly. Only where such regulation interferes with important family interests (such as access to special education or medical treatment that is unavailable in the employing community) would the residency requirement interfere with an individual's protected rights. (*Hameetman v. City of Chicago*)

Several state supreme courts also have sustained the validity of a local residency requirement. In *Gantz v. City of Detroit*, the Michigan Supreme Court upheld a Detroit Civil Service Commission's regulation requiring continuing residency for municipal employees. In 1973, the California Supreme Court sustained the City of Torrance's residency requirement.

Although a residency requirement may be ruled valid, questions may arise as to what constitutes "residency." In *Fagiano v. Police Board of the City of*

Chicago, the question of what constitutes "residence" was defined. In sustaining the city's residency requirement,[35] the Illinois Supreme Court held that "residence was intended to be synonymous with domicile, which has been defined as the place where a person lives and has his true, permanent home, to which, whenever he is absent, he has an intention of returning." The factors used to determine a person's actual residence or domicile constitutes another area of significant litigation.

For example, when a Detroit fire fighter's family lived outside the city and he lived within it, he was told that he did not meet the residency requirement unless his wife and family also lived in Detroit. Upon his refusal to comply with an order that they move back to the city, he was dismissed from his position. Although the circuit court held that he could not be considered a resident of the city while his family lived elsewhere, the Michigan Court of Appeals ruled otherwise. The appellate court held that it was an error for the lower court to rely solely on the residence of the plaintiff's family in determining his residence.

Other factors must be considered, for residency is essentially a matter of intent, which depends upon the facts and circumstances of each case. In *Civil Service Commission of Pittsburgh v. Parks*, it was found that the great majority of a fire fighter's living arrangements (location of family, receipt of mail, voting registration, vehicle registration, etc.) were at his residence outside the city, and that his apartment in the city was furnished but uninhabited. The court held that the fire fighter's residence was outside the city as a matter of law.

A state legislature may act to prohibit a municipality from passing a residency requirement, even though such a requirement is permissible under the U.S. Constitution. Such prohibitions generally have been sustained by the courts. In *Smith v. City of Newark*, the court examined a law passed by the New Jersey Legislature in 1972 which eliminated the residency requirement as a condition of employment or promotion of municipal fire fighters. This ruling superseded previous residency rules of municipalities, since a later law is deemed to repeal former rules where there is a conflict.[36] As a result, the

---

[35] "All officers and employees in the classified civil service of the City shall be actual residents of the City. Any officer or employee in the classified civil service of the City who shall fail to comply with the provisions of this section shall be discharged from the service of the City in the manner provided by law." Chicago Municipal Code Chapter 25, Section 30

[36] "No municipality shall pass any ordinance, resolution, rule, regulation, order or directive, making residency therein a condition of employment for the purpose of original appointment, continued employment, promotion, or for any other purpose for any member of a fire department and force and any such ordinance, resolution, rule, regulation, order or directive in existence on the effective date of this act or passed hereafter shall be void and have no force or effect." New Jersey Statutes Annotated 40A:14-9.1

plaintiff, a resident of Newark, was unsuccessful in obtaining the removal from the Captain's promotional eligibility list of names of 35 nonresident fire fighters who had placed higher than he did.

In sustaining the New Jersey statute, the court (in *Booth v. Town of Winslow*) reasoned: "The municipal ability to recruit the best available applicants may be severely constrained by a residency requirement, and obtaining the benefits of local residency may well result in a compromise in the quality of appointments resulting from a constricted recruitment pool. These are difficult conflicts to resolve. The Legislature has opted to resolve them by a uniform scheme, applicable on a state-wide basis, which permits municipalities to require residency as a condition of appointment so long as qualified resident-applicants are available but not as a condition of continued service or promotion." In sustaining a comparable law, in *City of Atlanta v. Myers* the Georgia Supreme Court reasoned that passage of such laws is intended to remove the chilling effect of mandatory residency, and permits a city to select those most qualified for public service.

An amendment to a state's constitution also can limit a fire department's authority to establish or maintain a residency requirement. In the state-wide election conducted in California on November 5, 1974, a constitutional amendment was adopted reading as follows: "A city or county, including any chartered city or chartered county, or public district may not require that its employees be residents of such city, county, or district; except that such employees may be required to reside within a reasonable and specific distance of their place of employment or other designated location."

This type of law overcomes the objection of employees who live in another community close to the city of their employment. One example is the case of Los Angeles city employees who live in Beverly Hills or any one of a number of communities surrounded by the City of Los Angeles. Where a fire fighter resides in a city covering hundreds of square miles, as in the case of Los Angeles, it is possible for him to live more than 50 miles from any station in the city to which he might be assigned. On the other hand, any fire fighter who lives 50 miles from his station in Washington, D.C. might be living in Maryland or Virginia. Whatever rule is adopted in defining a "reasonable distance," it probably should refer to geographical boundaries rather than a specific place of employment. This is important when applied to employees who can be transferred periodically from one location to another within a very large area.

No residency requirement may be discriminatory of individual employees nor in contravention with other state statutes. Where a state law permits removal of fire fighters only "for cause," removal of a one man due to lack of residency was held invalid. In this case, *White v. City of Winnfield Fire Department*, there was no showing that the fire fighter's residence outside the city had impaired the fire department's efficiency or fire protection capabilities.

Note that failure of a city to enforce its residency requirement upon an employee's initial appointment does not preclude it from enforcing the requirement later during the employment period. The rule on subsequent enforcement is aptly described in *City of Meadville Firemen's Civil Service Commission v. Neff*: "The fact that a municipality may not have enforced a residency requirement or rule in the past does not mean that it is forever precluded from attempting to enforce it at some future date. Any other result would be absurd. The municipality need only provide adequate or reasonable notice to those affected by the requirement that it will be enforced. . . ." Where a residency requirement was not published and recorded in the city's ordinance book, it was held unenforceable because the required notice was not provided to fire fighters. (*Brown v. City of Meridian*)

Eligibility Lists

Before an eligibility list is established, there usually is a period of several days in which to file a protest. If a candidate does not challenge the validity of the list within the prescribed period, in the absence of fraud or collusion he will be barred from any subsequent challenge. However, in one case a Fort Worth, Texas, fire fighter took an examination for driver in 1972 and placed 19th on the eligibility list. The exam was regraded by the Civil Service Commission. Even though the fire fighter had not filed a protest within the five days directed by the Commission, the reviewing court permitted suit and invalidated the revised list. It was held that the candidate had an equitable property right in the position that was worth enforcing, making it necessary to throw out the revised list. (*Crain v. Firemen's and Policemen's Civil Service Commission*)

Appointment of Fire Fighters

Aside from the problems of discrimination, there are other considerations when making appointments to the fire department. For example, where a state law of Indiana listed dispatchers and mechanics among the members of the "fire force," the court ruled that civilians appointed to replace fire fighters in these jobs must conform to the statutory requirements for fire fighters as to age and physical condition.

In a Kentucky case, the plaintiff sought to have the court direct the civil service examiners of the defendant city to certify his eligibility for appointment as a fire fighter. The defendants refused to certify him, instead certifying and appointing another candidate who had a lower grade. The court held that the plaintiff had a right to compel the defendants to comply with the ordinance and report the results of the examination, for this was a ministerial duty. However, he could not compel the defendants to certify him, as this was an act of discretion.

There is nothing objectionable in vesting in the civil service commission the power to remove names from an eligible list of candidates for appointment where the city charter provides for such action after the names have been on the list for two years. But a civil service commission cannot delegate its duty to determine the good moral character of candidates for appointment to the board of fire commissioners where the commission has been specifically directed to perform this determination.

Civil Service and the Probationary Period

Once an appointment has been made, a candidate may undergo a further test of his qualifications for the position—that of satisfactorily passing the probationary period prior to permanent appointment. Whether this period is for six months or a year depends upon the civil service rules applicable to any given position. *Fish v. McGann* stated: "The object of having a period of probation is to determine whether the conduct and capacity of the person appointed are satisfactory, notwithstanding the fact that he has passed the exam, and his appointment has been certified by the commission to the appointing office."

The probationary period has not been without legal challenge. Sometimes the evaluation following the probationary period was found to be clearly the result of bias and retaliatory discrimination; in other cases, the probationary period lacked training opportunities and constructive criticism of performance. Under such circumstances, the courts have overruled application of probation period results and have mandated permanent appointment of fire fighters. Just as the initial job entrance examinations and qualification requirements must be without inherent potential for job discrimination, so too must be the probationary period. (*Berkman v. City of New York*)

Rules and Regulations

Rules and regulations of fire departments usually are adopted to serve as a guide for conduct by the members of the department. A breach of these rules sometimes carries the penalty of dismissal. Where such is the case, state law requires that the accused be provided with the equivalent of due process—that is, notice of allegations and an opportunity to answer before an impartial tribunal.

Restriction to Filing a Lawsuit

A rule that requires a fire fighter to obtain permission from his superior officers or the board of fire commissioners to file a civil action is considered unreasonable, arbitrary, and unconstitutional. The Missouri case of *State ex rel. Kennedy v. Rammers* held that removal of a person based on a violation of this kind of rule or regulation was improper, taking away the constitutionally protected requirement that the courts of justice be open to every person, property, or character. It is also illegal for a department to "retaliate" against a member for assisting someone else to obtain the assistance of the courts. Where a father was effectively discharged from a volunteer fire department for assisting in his daughter's discrimination action against the department, he was entitled to full reinstatement with all applicable benefits, the court ruled in *Harmony Volunteer Fire Company v. Commonweath of Pennsylvania Human Relations Commussion.*

Grooming—Hair Styles, Beards

The right of fire departments to regulate the length of hair, mustaches, and sideburns, as well as to ban beards and goatees, has been contested unsuccessfully in a number of state and Federal courts. As stated in *Kelly v. Johnson*, adoption of such a regulation is permissible provided it bears a reasonable relationship to the needs of the department for the maintenance of discipline, esprit de corps, and uniformity. Under these circumstances, departments can freely establish requirements for grooming, even prohibiting beards, goatees, and/or mustaches. Where the department's reasoning is simply to maintain a military appearance, for the sake of uniformity, the rules also have been sustained. (*Muscare v. Quinn*) In addition, rules promulgated in the interests of fire fighter safety—such as to provide effective face seal for breathing apparatus—are being given greater latitude by the courts. The right and duty of fire department officials to achieve maximum safety is considered a major factor in determining reasonable regulations of fire fighter attire and grooming, it was ruled in *Michini v. Rizzo*.

Enforcement of grooming standards can be prohibited where they have not been lawfully promulgated. Where one board of fire commissioners failed to adopt a grooming rule, regulation, or bylaw, the fire chief was not empowered to enforce a grooming standard. *Ittig v. Huntington Manor* affirmed that only upon adoption of a grooming rule in conformity with due process standards can grooming requirements be put into effect and the rule enforced.

In a case where compliance with a grooming standard is medically unfeasible, disciplinary enforcement of the standard will most likely be prohibited. For example, in *Shelby Twp. Fire Department v. Shields* a Michigan fire fighter had a case of "pseudofolliculitis barbae," a condition caused by ingrown beard stubble. Shaving aggravates the condition, so the normal treatment involves cultivation of a beard—a growth prohibited by the fire department's grooming standard. The court held that the fire fighter was "handicapped" under Michigan law and could not be discharged due to his medical problem unless the fire department could prove he could not perform his job using adaptive devices or aids. Should no appropriate device be available, the fire fighter potentially could be entitled to a light duty position or release on a disability retirement.

"Moonlighting"—Outside Employment

The law relating to dual employment of public employees varies with the jurisdiction. In general, municipalities are entitled to restrict outside employment, especially where it could conflict with efficient fire department operations. As the Wisconsin Supreme Court stated in *Huhnke v. Wischer*: "It is conceded that the rule [against outside work] and the ordinance must bear a rational relationship to the maintenance of an efficient fire department. One of the reasons for the rule is to insure that members of the department will be at all times in physical condition to perform their duties as firemen if called upon to perform them. A fireman is subject to call for duty 24 hours a day; he is potentially on duty at all times during his off hours .... If the rule is harsh, it is for the individual to determine whether he will subject himself to its terms by becoming a member of the department."

Rules against "moonlighting" may be considered essential or at least appropriate in order for a fire chief to maintain control of a tightly disciplined group of employees. Their special duties and obligations to protect the safety and order of a city mean they must be available to cope with any emergency.[37] (*Calfapietra v. Walsh*)

Secondary employment restrictions can be applied retroactively. *Phillips v. Hall* found that a fire fighter does not have a property interest in his second job that prevents a department from prohibiting continuation of his outside work. *Gailband v. Christian* established that employees also can be terminated for performing unapproved outside employment during any period of sick leave, despite their inability to perform their department duties.[38]

Some fire fighters work on a 24-hour shift schedule, which often permits them to be off duty more than 20 days each month. Here the courts may be inclined to consider any rule that forbids outside employment as an unreasonable restriction. For example, the Supreme Court of Louisiana, in declaring the invalidity of the City of Crowley's ordinance forbidding "moonlighting" as applied to fire fighters, said: "Admittedly, were the firemen working under a differenthourly schedule which would result in a lack of available outside working time, this ordinance might well be reasonable in attempting to insure that firemen were alert, healthy and not lacking in sleep or rest due to outside employment." (*City of Crowley Firemen v. City of Crowley*)

Endorsing Products

A rule preventing use of a fire department member's name, picture, or image in any product advertisement or endorsement, without the approval of the fire chief, was held valid in *Kane v. Walsh*. Such a restriction can extend to prohibitions on outside business, especially where such undertakings present a potential conflict with fire department interests. One fire fighter's outside business involved the sale, installation, distribution, repair, and service of fire and smoke detectors. The fire department was judged within its au-

---

[37] This decision sustained the authority of the New York City Fire Department to restrict outside employment.

[38] A police officer was terminated for violating regulations, over an extended period, by performing unapproved work during a period of sick leave.

thority to prohibit his continued engagement in this business under penalty of termination. *Weisenritter v. Board of Fire and Police Commissioners of Burbank* said: "The possibility of conflict of interest is a reason for the prohibition against dual employment. It is difficult to conceive of a situation more prone to create such conflicts than that of a fireman moonlighting by selling fire alarm systems. His position as a fireman may lead persons he approaches to have confidence in the systems he is selling as well as to believe they might receive better fire protection by dealing with a member of the fire department." Such inferences obviously are contrary to the best interests of the fire department.

Off-Duty Conduct

A rule prohibiting off-duty conduct incompatible with a fire fighter's duties (unbecoming conduct) usually is held enforceable. In general, the conduct prohibited must directly interfere with fire department operations and efficiency. Proof of such interference must be presented if the case goes to court. However, not all jurisdictions require showing of proof. For example, in *Travis v. City of Memphis Civil Service Commission* the court found that a fire fighter's tacit approval, by his silence, of his wife's drug dealings from their home was conduct incompatible with the high moral standards required of fire fighters. Although the majority of jurisdictions would require a showing of how the drug sales actually affected fire department efficiency, none needed to be shown here. The fire fighter could be discharged simply for failing to distance himself from the felonious activity, ruled the court.

Sick Leave

Sick leave is a negotiated employee benefit which is not the subject of Federal or, typically, state regulation. Rather, it is a benefit obtained as a result of collective bargaining. Even though a person is entitled to the sick leave benefit, what restrictions may a department place upon its usage? May a department restrict an individual to his residence, except for medical consultation, during the term he is on leave? Can the department mandate periodic physical examinations throughout the period of sick leave?

During any sick leave period, a fire fighter still is subject to the directions of the fire department despite his release from active service. In *Driscoll v. Department of Fire of the City of Syracuse*, the plaintiff fire fighter was suspended for three weeks without pay for disobeying a direct order to undergo a medical examination. The court sustained the suspension, holding that refusal to obey the fire chief's order to attend the medical examination was a valid basis for suspension, even though the fire fighter claimed he "forgot." The court stated that it recognized that "the failure or refusal to obey an order of a superior officer is a serious matter," even during a period of sick leave. Mandatory attendance at a medical examination is within the authoritative province of the fire chief.

Department restrictions that prohibit a fire fighter from leaving his residence during a period of sick leave, except for medical reasons, generally are considered unenforceable. In *Pienta v. Village of Schaumburg, Illinois*, the court reviewed the city's sick leave residence restrictions. The court recognized the city's interests in avoiding abuse of its liberal sick leave policy and in protecting the public's resources, but held such restrictions unconstitutional. Where the individual and his family also were subject to surveillance inside and outside their home, frequent telephone calls, and unannounced department visitations, the court held such regulations infringed upon the family's rights to vote, right to free exercise of religion, and right to travel. "In effect, these regulations put plaintiffs under house arrest until their return to work. Their rights to vote, to exercise freely their religion by church attendance, to go to court, to attend political or family gatherings and to travel were infringed." Without extremely compelling municipal interests, such restrictions generally are impermissible because they limit individual freedoms, and in most cases are unenforceable as a matter of law.

Use or Possession of Liquor

The courts have shown no reluctance to uphold rules and regulations which forbid drinking alcoholic beverages while in fire department uniform or on duty. Rules which bar the possession or use of liquor on fire department premises are considered reasonable and not an unconstitutional deprivation of individual freedom. In addition, the courts have sustained regulations that prohibit a fire fighter from being under the influence of alcohol or drugs upon reporting for duty. The courts have reasoned that "the duties of a member of the city fire department are such as demand full possession of all his faculties at all times while so engaged, and there can be no question about the propriety of the involved rule in this case, since intoxication would render a member (and especially if his duties were to drive the fire apparatus over the streets of the city to the place of the fire) wholly inefficient, as well as dangerous to others using the streets, to say nothing about possible total dethronement so as to render the member wholly incapable in such an emergency." (*Northcutt v. Hardebeck*) Where a fire fighter was discovered to be "under the influence" following his failure to respond to an alarm call, his termination was deemed justified.[39] (*Nick v. Department of Fire*)

Being "under the influence" must be distinguished from the disability of alcoholism. Even though a fire fighter can be disciplined for reporting to duty while under the influence, when the condition has reached the state where the fire fighter can be classified as an alcoholic—a recognized disability—the

---

[39] A 60-day suspension and demotion of a fire captain was sustained despite the contention that his consumption of alcohol while on duty had no effect on his work performance, as evidenced by his satisfactory performance during a "roll" that evening.

department may not handle the situation simply with repeated discipline. Rather, a directive for rehabilitation, assignment to light duty during the rehab period, and termination only where rehabilitation is unsuccessful should be pursued. Until such opportunities have been offered and a fire fighter's total and permanent disability from alcoholism has been proven conclusively, dismissal of a fire fighter for repeatedly being under the influence would be prohibited. (*Zeigler v. Department of Fire*)

Partaking of Meals Together

A department requirement that all fire fighters take their meals together, thus prohibiting any fire fighter from leaving the station during the meal break, has been judged enforceable. As long as provisions of the rule are not arbitrary or capricious, and are in the interest of proper discipline and maintaining operational efficiency, they will be upheld in court. (*Warner v. City of Los Angeles*) One fire captain's unauthorized and unannounced absence of one hour and 45 minutes for a meal, during which time his unit was called for an emergency response, justified his 60-day suspension without pay, his demotion to the rank of fire fighter, and his transfer to another station. (*Newman v. Department of Fire*)

Membership in a Collective Bargaining Unit

Years ago it was held that "any association which, on any occasion, for any purpose, attempted to control the relations of members of either the police or fire departments toward the city they undertake to serve, is, in the very nature of things, inconsistent with the discipline which such employment imperatively requires, and therefore subversive of the public service and detrimental to the public welfare." This quotation is from the 1923 decision in *Hutchinson v. Magee* which held that a fire fighter who refuses to sever his connection with an association which had organized a strike of city fire fighters may be dismissed from the service. The court's statement reflects the historical perspective of fire service union membership prior to passage in 1935 of the National Labor Relations Act (usually called the Wagner Act) and state public employee relations acts.

Today, state and Federal courts which have considered public employee relations issues generally have sustained the fire fighter's right to join a union or other employee organization and to be represented collectively.[40] (*Town of Winchester v. Connecticut*) The organization, as the representing agency, thus is empowered to negotiate with the community over "terms and conditions of employment" including health- and safety-related issues. (*Town of*

---

[40] The sole paid member of the volunteer fire department has protected rights to belong to the union, but not to form a bargaining unit consisting of one individual.

Burlington v. Labor Relations Commission)[41] In addition, certain activities undertaken by a fire fighter, as a union representative, are considered constitutionally protected; this includes speaking to the press and communications outside the chain of command, for which department discipline is prohibited. In case of conflict, public employee relations laws permit the employee organization to seek corrective remedies from state agencies or employee relations boards without obtaining prior permission from the department. While the rights and privileges conferred upon collective bargaining entities are beyond the scope of this book, an understanding of some principal issues that affect fire department collective bargaining is worth mentioning at this point.

Public employee relations laws may prohibit "supervisory" personnel from participating within the same bargaining unit as fire fighters. Also, such laws may prohibit "managerial" personnel from participating in any collective bargaining unit. These laws are based upon the inherent conflicts among the interests of the respective groups and general employer-labor relations principles. Recurring questions under the state's public employee relations laws become: Which fire department personnel are considered supervisory personnel, prohibited from membership in the fire fighters' union? Which are managerial personnel, prohibited from any collective bargaining unit membership? No universal rule can be formulated, "whether employees in a particular case are managerial [or supervisory] is a factual matter to be determined by an examination of their duties and responsibilities [in each case]." (*City of St. Augustine v. Professional Fire Fighters*)[42]

Are issues of apparatus or fire scene minimum staffing a "term or condition of employment" subject to mandatory collective bargaining negotiation? Again, there is no uniform rule. The primary question is whether or not the minimum staffing is truly an issue of safety. Is a minimum number of staff necessary for safe operation of a particular piece of apparatus or equipment or at a fire scene? Is there a managerial staffing question, such as the total number of fire fighters to be on-shift at any time? Where the issue is truly a safety question, courts have considered the issues to be subject to collective bargaining. (*IAFF v. City of Salem* and *City of Erie v. IAFF*) However, where

---

[41] "It is settled that promotional opportunities constitute terms and conditions of employment and that a denial of a promotion based upon an employee's union activity is a prohibited practice. . . ." This meant that a deputy fire chief could not be passed over for promotion to fire chief because of his union activities.

[42] Duties and operating procedures of the fire department brought captains clearly within the statutory definition of "managerial employees."

A contrasting case within the same state (Florida): *City of Jacksonville v. Jacksonville Association of Fire Fighters*. Captains are *non*-managerial.*Ocean City-Wright Fire Control District v. Ocean City-Wright Fire Fighters Association, Local 2879, IAFF*. Fire captains and inspectors *can* be in fire fighters' collective bargaining unit.

the issue is the minimum number of fire fighters who must be on duty at any one time, and not simply the number who must be available to operate a piece of equipment when needed, such decisions generally are held to be exclusively managerial in nature, not subject to mandatory collective bargaining. (*City of New Rochelle v. Crowley*)

Financial Disclosure

Where adequate and effective privacy protections exist, the courts generally have sustained a community's right to require public officers and employees to disclose information on their personal financial position and their ownership of real estate within the community. Reasons for such disclosure have included enhancement of public confidence in the integrity of city government and the avoidance of conflicts of interest or even of any inference of such conflict.[43] (*Barry v. City of New York*) In one case, several newspaper articles indicated that certain city employees owned properties which were in violation of city housing codes. The potential for preferential treatment by city inspecting authorities was inferred. As a result, the city was justified in mandating the public disclosure by fire fighters of all real estate they owned within the city. (*Evangelista v. City of Rochester*)

Rules and Regulations Affecting Speech

What effect can a fire department regulation have upon a fire fighter's constitutionally protected right of free speech? Generally, every public employee is fairly free to express his views, in public or private, orally or in writing. (*Abood v. Detroit*) However, where the expression affects the employee's position or department, it is subject to greater judicial scrutiny.

Waters v. Chaffin says: "Despite the fact that public employees do not relinquish the protections of the First Amendment by accepting a job with the government, their rights are somewhat less extensive than those shared by the citizenry at large.... In any particular case when a public employee speaks out, then, it is necessary to strike the proper balance between the interest of that employee, [as a citizen], in speaking freely [upon matters of public concern] and the interests of the state, as an employer, in promoting the efficient delivery of public services [it performs through its employees]." With regard to fire protection, the courts have held the high stakes of fire fighting necessitate maximum operational efficiency. Therefore, courts weigh the state's interest in regulating speech of fire fighters more heavily than the general public employee. (*Bickel v. Burkhart*)

Factors a court considers in determining the appropriate balance include:

---

[43] New York City's financial disclosure law, reguiring all fire department personnel earning $30,000 or more annually to file an annual financial disclosure statement, was upheld.

1. The need for harmony in the office or work place;
2. Whether the government's responsibilities require a close working relationship to exist between the plaintiff and co-workers when the speech in question has caused or could cause the relationship to deteriorate;
3. The time, manner, and place of the speech;
4. The context in which the dispute arose;
5. The degree of public interest in the speech; and
6. Whether the speech impeded the employee's ability to perform his or her duties. (*Bowman v. Pulaski County*)

In the contexts that follow, department regulations affecting free speech are discussed along with the extent to which the courts permit their enforcement.

Going Through "Channels"

A rule that requires fire fighters to report all activities or complaints through their superior officers generally is enforceable. When disputes arise, the courts typically examine the circumstances of the case to determine the importance of enforcing chain-of-command requirements. "We cannot simply decide in the abstract. . .that a chain-of-command policy designed to protect this interest [that is, department loyalty, discipline, and uniformity of operations] will always take precedence over the interest of a public employee in open communication. Rather, we must look to the particular circumstances of each case to determine the importance of enforcing the chain-of-command [requirement] against an employee whose speech breaches that policy." (*Brockell v. Norton*)

When the issue involves a grievance or the imposition of discipline, a "through channels" requirement generally will be sustained. A California court sustained a chief's termination order concerning a fire fighter who had been suspended for wearing nonuniform shoes. The fire fighter persisted in taking his grievance—loss of a day's pay—to the city administrative officer, in direct disobedience to an order from the chief that he must process his complaint through proper channels.[44]

Where the reporting involves notification of an illegal act or activity (that is, "whistleblowing"), the courts generally hold that "through channels" reporting is not enforceable. *Brockell v. Norton* ruled that "An employee's First Amendment interest is entitled to more weight where he is acting as a whistleblower exposing government corruption." Where a private call was made to an authority positioned to make a quick investigation and resolution of al-

[44] Morse, H. Newcomb, "Legal Insight: Willful Disobedience," *Fire Command*, Vol. 39, No. 11, Nov. 1972, p. 36

leged improprieties, the court held that such action was less disruptive of department morale, so the chain-of-command rule could not be enforced.

Criticizing Department Policy

In addressing issues of critical speech, the rules distinguish between public and private expression. Where the expression is public, it is generally the content of the statements that must be assessed to determine whether or not the speech is protected. Where private expression is involved, the balancing test focuses not only on the content of the employee's speech—the main focus when the employee voices his criticisms publicly—but also on the manner, time, and place in which it was delivered. (*Bickell v. Burkhart*) For example, critical commentary during a private conversation between supervisor and employee would be protected, but subsequent expression of dissatisfaction to others is not protected. (*Givhan v. Western*)

For example, a fire fighter made statements of criticism, at a department meeting where critical comments could reasonably be expected. The comments, aimed at the department as an institution, and not directed at any individual member, were held constitutionally protected. Because the comments were not made in a hostile or threatening manner, nor was there any indication that the statements would interfere with operations, the department's interest in maintaining discipline and efficiency were not affected.

Applying the same time, place, and manner considerations in *Janusaitis v. Middlebury Volunteer Fire Department*, the court reached a contrary conclusion. In that case, the fire fighter engaged in threats, "carping criticism and abrasive conduct." In addition, the trial court found that the fire fighter was motivated by deep resentment and intense bitterness rather than by a concern for the welfare and efficiency of the fire department. Thus the court reasoned that such speech could undermine efficiency, justifing departmental discipline.

In another instance, *Germann v. City of Kansas City*, a fire captain used a circulated letter to accuse the fire chief of being a "liar," of tearing the department to shreds, of having a "pitifully twisted outlook," and of obstructing the department more completely than anyone on the face of the earth. The court in this case decided that the chief was justified in concluding that the captain would not support the chief's policies if he were promoted to battalion chief. Although the captain was an officer in the fire fighters' union at the time of the incident, his speech was not protected as part of constitutionally protected right to engage in union activities. The fire chief was deemed justified in selecting someone else for promotion to the battalion chief's position.

In contrast, where a police officer called his chief a "bastard" and a "son of a bitch," these comments were held protected as they were uttered in a private conversation, while out of uniform, out of the department's jurisdiction, and were made to a person considered a friend. In *Waters v. Chaffing*, the court held that where the department could make no showing of actual harm or reasonable likelihood of harm to efficient operations, such speech was protected.

Another case involved a fire fighter who called the shift commander a "big dummy" and all other fire fighters on duty "dummies." The comments were considered part of general fire department banter, not an expression of disrespect, and thus were not justifiable grounds for the fire fighter's dismissal. (*Hutchinson v. Hughart*)

The concept of protected free speech is not limited to the printed or spoken word. Physical representations and gestures, arm bands, buttons, and other items can be considered expressions of speech. For example, where a fire department instituted disciplinary proceedings against fire fighters for wearing their uniforms during publicity functions, the union sought a court injunction. The court stated: "The real issue boils down to the question of whether the wearing of the uniform comes within the free speech protection of the First Amendment. Does it embody the message plaintiffs were attempting to convey in a manner that can be likened to the black armbands which the [United States] Supreme Court found 'protected' in *Tinker v. Des Moines Community School District*." Where wearing of uniforms had no "independent significance," it would not be protected speech. However, should the items worn independently express the views of the fire fighters (such as black armbands or protest buttons), the items may be considered constitutionally protected expressions of speech.

Courts generally balance the interests of the city in maintaining the appearance of the uniform as politically neutral with interests of the fire fighter in expressing his position. Then the courts determine whether or not the nonverbal expression should be protected. (*Detroit Fire Fighters v. City of Detroit*)

Speaking to the Media

A rule prohibiting fire fighters, whether on or off duty, from speaking—favorably or not—to the media on any subject pertaining to department activities has been held unconstitutional. Such an order constitutes "a vague and overbroad regulation of speech, regardless of content or effect, [and has a] substantial chilling effect upon the exercise of First Amendment rights." The prohibition of any and all speech before it occurs is characterized as an unconstitutional prior restraint and as an [impermissible] "gag rule." Such prior restraint on individual free speech violates the First Amendment except in the most compelling circumstances. (*Grady v. Blair*) Discipline proceedings for violating such orders typically are found without merit and are reversed.[45] (*Klein v. Civil Service Commission*)

---

[45] A fire fighter's suspension for releasing news of a labor dispute, in violation of an order not to release news of departmental affairs without the chief's prior approval, was reversed.

Engaging in Political Activities

A rule that totally forbids fire fighters from taking an active part in politics or engaging in controversy concerning candidates and issues is unconstitutional. However, courts differ in defining the manner and extent of fire fighter participation. For example, where a fire fighter sought election to his city council, in violation of a municipal ordinance prohibiting such dual position holding, the court held that public employees may become candidates for public office and may speak freely on public issues and controversies. No one charged the fire fighter with conducting his campaign while on duty, or with neglecting his job to achieve his political objectives.[46]

An expanding minority of jurisdictions, however, permit government to completely ban dual position holding. "The purposes attributed to such provisions include: 1. [preventing] multiple position-holding, so that offices and positions of public trust would not accumulate in a single person; 2. preventing individuals from deriving, directly or indirectly, any pecuniary benefit by virtue of their dual position-holding; 3. avoiding the inherent conflict which occurs when an employee's elected position has supervisory power over the employee's superior in another position; and 4. generally insuring that public officeholders and public employees discharge their duties with undivided loyalty."[47] (*Acevedo v. City of North Pole*)

Between these extremes, the courts tend to examine any particular activity, determine whether or not it is political in nature, and decide if the community has a compelling reason to prohibit it. Activities deemed partisan can be subject to relatively greater limitations, even outright prohibition, whereas nonpartisan activities are subject to very limited local regulation.[48] (*Magill v. Lynch*) This is because "government and its employees in fact avoid practicing political justice, but it is also critical that they appear to the public to be avoiding it, if confidence in the system of representative government is not to be eroded to a disastrous extent." (*U.S. Civil Service Commission v. National Association of Letter Carriers*)

For activities in support of partisan politics, the courts have sustained restrictions upon a fire fighter's authority to publicly endorse a candidate, circulate petitions, participate in public rallies, contribute to or solicit contributions for council candidates, or act as a campaign manager. Partisan activities that have been sustained include placing campaign signs in yards, wearing candidate buttons, affixing bumper stickers, and permitting spouses to contribute and solicit contributions for candidates. However, even other-

[46] Morse, "Legal Insight: Candidate for Public Office," *Fire Command*, Vol. 36, No. 2, Feb. 1969, p. 45

[47] Court upheld the firing of a police officer upon his election to the city council.

[48] Government may constitutionally restrict Pawtucket, Rhode Island, fire fighters' participation in nominally nonpartisan elections if political parties play a large role in the campaigns.

wise permissible activities may not be conducted at the workplace or at any location where they would directly interfere with department functions. (*Wachmans v. City of Dallas*)

Courts generally prohibit department restrictions on any activity connected with a truly nonpartisan election. Examples might be an initiative to obtain pay raises, or a bond issue for new equipment. Where fire fighters circulated a petition in an attempt to get passage of an ordinance that would reduce their working hours, the Alabama Supreme Court held there was no violation of a department rule prohibiting "taking part in a political campaign."[49] It is not how the election is publicly classified that defines the permissible restrictions; rather, it is the actual, factual circumstances of the election. Where a municipal ordinance restricted all political activities, partisan and nonpartisan, the court held that such prohibitions were unconstitutionally overbroad infringements of the franchise right. (*Arden v. Village of Oak Lawn*)

Review Activities

1. The early colonial law that imposed a fine of 15 shillings on "every person who shall suffer his house to be on fire" was probably the earliest American legislation relating to fire safety.
   - A. Explain why you feel this was or was not a good law.
   - B. Do you think that if fees are imposed for rendering fire suppression services, the increasing number of building fires might be reduced? Why or why not?
   - C. For the catastrophic fire incident, should the property owner be liable for the department's costs for rendering services? How would you calculate the costs?
   - D. Explain whether or not you feel a fine or other punishment should be imposed on the owners of buildings that have caught fire.
2. Outline what you feel would be a successful fire department plan for carrying out a lot cleaning program. Include in your outline descriptions of some of the legal problems you might encounter, such as determining ownership, giving appropriate notice, entry upon property, seizure and disposal of refuse, charges and compensation for fire department costs, and contracting for services by an independent firm. What actions would you recommend to address these issues?
3. With a group of your classmates, discuss whether they feel a fire fighter is a public officer or an employee. Then interview some residents and

---

[49] Morse, "Legal Insight: Can Firemen Circulate a Petition for Signatures?" *Fire Command*, Vol. 39, No. 6, June 1972, p. 32

some fire fighters from your community to determine: 1. in which of these ways fire fighters are viewed; and 2. how fire fighters view their own role. Compare your findings with those of your classmates.

4. Explain why each of the following statements is true or false:
   A. The fire chief is head of the fire department, and the department's board of directors comes under the fire chief's jurisdiction.
   B. Workers' compensation only covers the injuries of a fire fighter occurring during a fire, not those of a volunteer who was assisting at the fire fighter's request.
   C. Volunteer fire fighters are not entitled to workmen's compensation if injured while traveling home from a fire.
   D. A volunteer fire department is not subject to the requirements of the civil rights laws because of its status as a volunteer organization.
   E. A fire fighter may not be constitutionally prohibited from holding both a fire department and an elected city council position.
   F. All fire fighters are "peace officers" and thus can carry concealed weapons.
5. As your department's public information officer, you wish to present a positive and appropriate public image of the department. Numerous incorrect and reputation-damaging statements have been issued by department personnel at fire scenes. What actions and directives would you recommend to remedy such a situation?

Chapter Six

A City's Liability for Acts of Its Fire Department

Tort Liability

The liability of a city toward the general public is governed principally by the laws of tort liability. A "tort" is an injury or wrong committed, either with or without force, to another person or his property; such injury can arise by the nonfeasance, misfeasance, or malfeasance of the wrongdoer. "Nonfeasance" is the failure to perform a duty, such as one owed to the public; "misfeasance" is the performance of an act which a person ought not to do at all; and "malfeasance" is the improper performance of an act which a person might lawfully do.

For a court to determine that a tort has occurred, the court specifically must find the existence of all of the following:

1. An act or an actionable omission by the defendant.
2. A duty of due care. Each person owes a duty to conduct himself as the average reasonable person would have done under the same or similar circumstances.
3. Breach of duty. Breach of the required duty of care owed the plaintiff.
4. Causation. The breach of the duty owed the plaintiff actually (either directly or indirectly) and proximately (close in time, chain of events, activity, duration, etc.) caused the plaintiff's injury. The cause, in a natural and continuous sequence of events, unbroken by any intervening cause, produced injury; without the cause, the result would not have occurred.
5. Damages. The plaintiff actually has sustained or will, in the immediate future, sustain damages as a result of the defendant's tortious (wrongful) actions.

Of the five elements required to establish municipal or fire department liability, determination of the existence of a "duty" to the plaintiff historically has been the decisive one. Without the presence of a legal duty, there can be no liability. It also is important to remember that the concept of duty is not firmly defined or limited in the law. Rather, it should be understood that duty is but "an expression of the sum total of those considerations of policy which lead the law to say that the particular plaintiff is entitled to protection," as stated by Prosser in *Handbook of the Law of Torts*.[1]

Background of Municipal Liability

The liability of a municipal corporation for its tortious acts, and those of its agents and employees, has been described as an extremely baffling concept. This confusion stems from the clashing theories of law fundamental to the subject, principally those theories describing the rights of individuals and those outlining the authorities and duties of government. The shifting of emphasis among these theories, either in support of or against a particular philosophy, has resulted in the myriad rules governing municipal liability. Also, as the subject of municipal liability is principally governed by state law, as many as 50 variations on the rules of liability can exist.

Municipal liability can be characterized as continuous conflict among the following legal principles:

1. "The king can do no wrong" or the "sovereign immunity" concept. As most rights and powers of individuals, townships, cities, and counties emanate from the sovereign authority of the state, specific authority to sue the state and its subdivisions must be granted by the state.
2. An individual has the right to seek redress of injuries resulting from socially unreasonable conduct.
3. Permitting individual recovery against the state will deplete the government's limited resources to rectify hazardous conditions for the greater benefit of all.
4. The fundamental rights of individuals should not be abridged without a compelling governmental interest.
5. The rights of the individual should yield to the authority of the government for the greater benefit of all.
6. The master should be held responsible for the actions of his servants when they are acting at his direction or in his behalf.
7. Where government is performing a service that also can be performed by a private entity, the rules of liability of private entities should apply.

[1] Prosser, William, L., *Handbook of the Law of Torts*, West Pub. Co., 4th Edition, 1971, p. 325

In Chapter 2, the topic of what is a tort was discussed. That chapter should be reviewed at this time, as the material that follows will focus on how the courts have tended to analyze the question of duty in the fire service setting.

History of Fire Department Liability

For centuries, it was a well-established principle of law that fire fighting was a governmental undertaking—that is, an undertaking that could be performed effectively only by an agency of government. The intended benefit conferred by the fire department was for the community as a whole (such as to prevent a conflagration) and not for the protection of individual interests. Therefore, the duty owed by the fire department was considered limited, restricted to protecting the total community. Thus, in performing a total community or governmental function, the fire department was immune from liability for the losses of an individual, whether occurring by acts of omission or commission, misfeasance or nonfeasance.

That concept of "A duty to all is a duty to no one" received extensive critical commentary. For instance:

"We think that a sound public policy requires that public officers and employees shall be held accountable for their negligent acts in the performance of their official duties to those who suffer injury by reason of their misconduct. Public office or employment should not be made a shield to protect careless public officials from the consequences of their misfeasances in the performance of their public duties," according to *Florio v. Schmolz*.

The inequities of "the king can do no wrong" concept resulted in the judicial formulation of numerous exceptions to the general rule of nonliability. While in the U.S. there had been practically a universal rule of nonliability of a municipality for torts with respect to fire suppression activity, the immunity became qualified for those activities not considered essential to the effective performance of fire suppression.

Examples of exceptions to the general immunity rule included construction and maintenance of fire department facilities, repair of fire alarm systems, and operation of fire department vehicles. The key to determination of the existence of immunity from liability focused upon whether the operation required the exercise of professional judgment, or was simply the execution of established policy, not requiring exercise of judgment. The question of municipal liability tended to fall into the following classifications:

1. For those municipal activities which fell in the category of proprietary functions (that is, not requiring exercise of judgment), no immunity existed. A municipality had the same tort liability as a private corporation.
2. For those activities which fell in the category of governmental functions, a municipality was liable in tort under the doctrine of *respondeat*

superior (meaning that the municipality is responsible in certain cases for the wrongful acts of its employees). This was true only when such tort was committed against a person with whom the agent or employee was in privity, or had induced detrimental reliance.
3. For those activities which fell in the category of judicial, quasi judicial, legislative, or quasi legislative, characterized as requiring the exercise of governmental judgment, a municipality remained immune. (*Commercial Carrier Corp. v. Indian River County*)

Historical Liability Principles in the Fire Service Setting

Examination of cases of various jurisdictions which have considered the historical question of municipal liability for fire department operations has revealed the general reasons leading to the conclusions attained. These reasons can be classified as follows:

1. A municipality may not be required by the legislature to establish and maintain a fire department. With no statutory obligation to establish a fire department, there is no duty to protect the public or any individual's person or property.
2. Where a community is obligated to establish a fire department, that department need not be capable of controlling all fire emergencies. Certain judgmental decisions (such as the level of fire service) should remain exclusively at the discretion of governmental authorities. "To accept a jury's verdict as to the resonableness and safety of a plan of governmental services and prefer it over the judgment of the governmental body which originally considered and passed on the matter would be to obstruct normal governmental operations and to place in inexpert hands what the legislature had seen fit to entrust to experts." (*Weiss v. Fote*)
3. The government's discretionary decision to establish a fire department should not establish liability for failure to provide fire suppression services. What has been provided by free will may be removed, as long as no detrimental reliance on the part of an individual has occurred. "To hold otherwise might well frighten our municipal corporations from assuming the startling risk involved in the effort to protect themselves from fire." (*Long v. Birmingham*)
4. Sovereign authorities ought to be left free to exercise their discretion and choose the tactics (such as the means of fire suppression) deemed appropriate, without worry over possible allegations of negligence. (*Wong v. City of Miami*)

5. Without a contractual obligation to provide a service, there is no duty to perform it.

However, even with these reasons, public dissatisfaction with the liability law continued. Further exceptions to the rule of nonliability developed. The result was general abolition of the universal rule of immunity, with the question of liability dependent upon the actual nature of the fire department activity.

Characterizing Fire Service Operations as Governmental v. Proprietary or Discretionary v. Ministerial

With abolition of sovereign immunity, most states retained qualified immunity for cities' governmental activities. These are activities requiring governmental decision-making of a high order—determinations requiring exercise of professional judgment on the part of government officials. This is contrasted with cities' proprietary activities, which do not require official exercise of judgment and are characterized as merely executionary in nature. Thus, "In determining whether an act falls within the exception [to the rule of liability], the court must consider the following four questions: 1. Does the challenged act, omission, or decision necessarily involve a basic governmental policy, program, or objective? 2. Is the questioned act, omission, or decision essential to the realization or accomplishment of that policy, program, or objective as opposed to one which would not change the course or direction of the policy, program, or objective? 3. Does the act, omission, or decision require the exercise of basic policy evaluation, judgment, and expertise on the part of the governmental agency involved? 4. Does the governmental agency involved possess the requisite constitutional, statutory, or lawful authority and duty to do or make the challenged act, omission, or decision? (*Evangelical United Brethren Church v. State*)

The fact that an employee normally engages in discretionary activity is irrelevant if, in a given case, the employee did not render a considered decision. (*King v. City of Seattle*) Thus, a simple decision whether to dispatch the fire department to the scene of a fire, or merely to investigate a smoke condition, does not involve a basic policy decision. That simple decision would be considered ministerial, for which no immunity would apply. However, sometimes the decision involves strategic or tactical considerations, such as how specific fire suppression forces are utilized. Such a decision would be described as judgmental and thus would be considered immune from negligence liability.

Retention of limited sovereign immunity based upon the characterization of the activity—governmental versus proprietary, or discretionary versus ministerial—is the rule in many states. However, other states have abolished sovereign immunity in favor of statutory protections for specific government activities. These protections typically are called statutory immunities.

Statutory Immunity

"Statutory immunity" is the granting by the state of an exemption from the performance of some duty. In the alternative, it is the granting of an exemption from liability for failure to perform or negligent performance of some duty which the law generally requires other citizens or entities to perform. Where sovereign immunity has been abrogated, a statutory immunity law effectively precludes liability, but only for those activities specifically enumerated in the immunity statute.

An example of statutory immunity is California's Government Code Section 850.4 which provides: "Neither a *public entity* nor a *public employee*, acting in the scope of his employment, is liable for any injury resulting from the condition of fire protection or fire fighting equipment or facilities or for any injury caused in fire fighting except for negligent operation of a motor vehicle." Thus, although sovereign immunity has been abrogated in California, the legislature has granted local government and its fire fighting employees statutory protection in the performance of fire suppression services. The reasoning for adoption of statutory immunity was "... adequate incentives to careful maintenance of fire equipment [exist] without imposing tort liability; and firemen should not be deterred from any action they may desire to take in combatting fires for fear that liability might be imposed if a jury believes such action to be unreasonable."[2]

Statutory immunities generally are narrowly construed. In interpreting statutory immunities, courts follow the rule of construction that furthers the goal of compensating injured parties for damage caused by negligent acts. The courts tend to bar government liability only where the legislature clearly has intended immunity. (*Potter v. City of Oceanside*)

In reviewing the scope of a community's or an individual's statutory immunity, extreme care must be taken to determine what activities are covered. The California Court of Appeals in its review of Section 850.4 noted that the immunity was intended to cover only negligent acts committed during the course of fire fighting. In *Potter v. City of Oceanside*, cited above, the injured plaintiff was operating a trenching machine which punctured a natural gas pipeline, causing gas to escape. When the city's fire department responded to the plaintiff's request for help, the fire captain negligently told the plaintiff to start the trenching machine and move it to another area. When the plaintiff did so, the escaping gas ignited and exploded, injuring the plaintiff. The court of appeals held that the immunity statute barred liability only for injuries caused in fighting fires. As the negligent act which resulted in injury occurred before any fire started, the city could be held liable.

In *Lewis v. Mendocino Fire Protection District*, the fire department was called to rescue a victim trapped in his tent when a tree fell on it. Holding the

---

[2] California Law Revision Commission Report, 1963, p. 862

fire district potentially liable, the court noted: "The legislature could have exempted defendant from tort liability arising from all emergency rescue operations, and not merely those related to combatting fire. The legislature did not do so. We commend the volunteers of the defendant fire protection district for their selfless and heroic acts in the face of substantial risks to their own safety. We are also aware that our decision might have a chilling effect on a fire department's willingness to respond to life-threatening but non-fire related emergencies. However, to construe the statute as defendant [fire protection district] requests would usurp the policy-making function of the legislature. The legislature is better equipped to strike a balance between the need to encourage these noble activities of rescuing people in emergencies and the need to provide compensation to the person injured by the negligent conduct of public employees." Statutory immunity was not applicable, as no fire fighting was involved.

In summary, statutory immunities granted by state legislatures are strictly interpreted and applied. Although courts recognize the public benefits granted by fire departments, the courts are powerless to relieve fire departments from liability where an immunity statute is silent on the subject.

Special Duty Doctrine

Although a governmental entity is immune from liability for truly discretionary undertakings and determinations, that exemption may be voided where a special relationship has developed between the defendant fire department and the injured plaintiff. Typically, this occurs when the fire department undertakes a discretionary activity; induces reliance by the plaintiff as to the continued or complete performance of that activity where it is reasonable for the plaintiff to rely on such assurances; and the department subsequently ceases performance without notifying the plaintiff.

Keys to determination of the special duty relationship are analysis of the reasonableness of a plaintiff's reliance on the fire authority's continued performance; the gravity of potential injury; and actions required of the fire authority that would have precluded the injury. The courts, in finding that a special relationship exists, have focused on the following:

1. The individuals to receive the benefit of the fire authority's actions are specifically identified and have direct contact. The individuals must be readily distinguishable from the public as a whole. (*Helman v. County of Warren*)
2. The fire authority was in a position of superior knowledge or authority, which could reasonably induce justifiable reliance by a plaintiff on the statements of the defendant.

3. The extent of misfeasance must exceed the concept of mere negligence. (*City of Atlanta v. Roberts*) Thus, where a city's fire safety code required the city to perform specific acts upon finding that a hazard existed, failure to perform could be considered gross negligence. (*Campbell v. City of Bellevue, Washington*)
4. The fire authority communicated its assumption of the corrective action obligation to the plaintiff, either by words or by actions to establish a special duty to protect the plaintiff. (*Helman v. County of Warren*)

In summary, a governmental unit will be considered to owe a particular individual a special duty of care when its officers or agents, in a position and with authority to act, have or should have had knowledge of a condition that violates safety standards prescribed by statute or regulation and that presents risk of serious harm to the specific individual or his property. Thus, when serious injury is reasonably foreseeable, the governmental unit has a duty to exercise reasonable care for the specific individual's benefit. (*Lorschbough v. Twp. of Buzzle, Minnesota*)

Summary of Municipal Liability Today

The conflict between the legal principles and resulting rules of law continues. However, the historical protections afforded by sovereign immunity, which served to delineate the existence of municipal immunity or liability, are gone. The general rule today is liability; immunity is the exception. Thus, except for activities protected by statutory immunity, if the following questions are answered in the affirmative, municipal liability generally will be present:

1. Was the duty violated connected with a proprietary rather than a governmental activity, or with a ministerial rather than a discretionary function?
2. Was the negligent person a servant of the city?
3. Was the servant's act *ultra vires* (beyond a person's scope of responsibility and authority)?
4. If the servant was not acting within the scope of his authority, was his act subsequently ratified by the municipality?
5. Did the city have a duty to protect the individual? If not, did the victim reasonably and detrimentally rely upon the affirmative representations of the city, its agents, or employees?
6. Was the municipality guilty of negligence, and was the plaintiff free from contributory negligence? If the plaintiff is an employee of the city, is he precluded from recovering therefrom by the assumption of the

risk doctrine, the fellow servant rule, or provisions of workers' compensation law?

The sections that follow show how the fundamental principles outlined above have been applied to various functional areas of the fire service. Throughout the discussions, it is important to keep in mind the legal concepts discussed. Particularly, was a governmental versus proprietary/discretionary versus ministerial function involved? Did a special relationship exist? Was statutory immunity applicable? Through understanding the basic legal principles and their application, an appreciation of the fire department liability exposure can be achieved.

Fire Suppression Activities

For many years, it mattered little how inefficient or inattentive a fire fighter was, or how poorly the fire department fought a fire. Performance of fire suppression services was considered a governmental duty—a duty owed the public as a whole, not a duty for benefit of individuals. The sweeping blanket of governmental immunity served to protect the fire service from critical examination. Now, however, the continuing trend of the respective states is to limit immunity, by finding that the fire service is a ministerial function or has a duty to specific individuals where no special relationship previously existed. This trend has served to expand the fire service's exposure to liability.

The principal avenue used for expanding fire suppression liability has been by finding the existence of a special relationship (or special duty between the fire department and the plaintiff.) A leading case on the subject is *City of Fairbanks v. Schiable*—an action brought against the city for the death from asphyxiation of a tenant of an apartment building. The suit charged negligence with respect to fire fighting. A woman, Druska Schiable, died of asphyxia during a fire in the building where she and her husband had an apartment. As her executor, her husband brought a wrongful death action against the landlord and the City of Fairbanks. The trial court rendered a judgment against both, and the case was appealed to the Supreme Court of Alaska.

The facts of the Schiable case were that the building was a four-story, concrete structure, containing a dumbwaiter used for carrying garbage cans from upper floors to the basement. Ventilating ducts interconnected all the kitchens, bathrooms, and clothes closets of all the apartments. The fire, started by children near a dumbwaiter door on the second floor, quickly spread to the third floor where the Schiables resided. Burning was intense along the wooden wainscoting on the stairway and in the hallway, creating a great amount of smoke. Heat melted the soldered ventilation ducts, and the accumulated grease burned. The decedent, unable to escape via the hallways, was seen at an exterior window awaiting rescue.

On the question of municipal liability, the Alaska Supreme Court made the following preliminary observation as to the applicable rule of law:

"Except for the negligent operation of fire equipment on public streets, where liability has been imposed on the theory of nuisance, it appears to be the general rule without exception that a fire department maintained by a municipal corporation belongs to the public or to the governmental branch of the municipality, and that the municipality is not liable for injuries to persons or property resulting from negligence with the fire department's operation or maintenance."

However, the state supreme court sustained the trial court's finding that a special relationship existed between the decedent and the fire department. It held that the fire department's failure to rescue the decedent, and its affirmative action in preventing others from saving her, amounted to negligence that was the proximate cause of her death. Then the court elaborated on its reasoning as follows:

"With her escape through the hallway cut off by heat and smoke, she was standing by her open window awaiting rescue. A fire fighter had climbed up a 24-foot ladder he had placed against the building in an attempt to reach her, but it was 8 to 10 feet too short. Efforts were made to place a 12-foot ladder on top of the 24-foot one, but this was found impractical. The fire department had one 50-foot ladder but it was being used to evacuate tenants on the other side of the building.

"We do not find that the lack of an additional ladder of sufficient length alone would be sufficient evidence upon which to base a finding of negligence; for other circumstances show that the fire department stepped out of its traditional role of simply doing the best it could to fight the fire and save lives. It assumed a particular obligation to save Mrs. Schiable's life, and then failed to fulfill that obligation by not using other available means to effect her rescue.

"While a fireman was backing down the ladder after an unsuccessful attempt to reach the decedent's window, he said to her: 'Don't jump, we'll get you out.' Efforts were then made to find a longer ladder but the taller ladders were all in use elsewhere. Meanwhile, no attempt was made to rescue the decedent with a fire net. Also, no attempt was made to place an available fire truck against the building and place a 24-foot ladder on top of it in order to gain the additional few feet needed to reach the decedent's window. From ladders too short to reach other apartment windows, ropes had been thrown to other persons awaiting rescue; but that wasn't done for Mrs. Schiable. The fireman who reached the top of the 24-foot ladder, and was only 8 to 10 feet from the window, made no attempt to deliver a rope to Mrs. Schiable. In fact, he couldn't remember whether there was rope on the rig or not.

"The city was guilty not only of failure to use common-sense methods to rescue the decedent, but also of affirmatively preventing rescue by others. Some spectators obtained an extension ladder of sufficient length to reach the decedent's window. They raised the ladder and were starting to extend it when they were ordered by a city fire inspector to get away from the building. When they refused to obey, they were driven off by fire hoses turned on them by order of the fire inspector.

"A bystander named Anderson climbed a ladder, put his head and shoulders in the window, grabbed Mrs. Schiable by the foot and attempted to pull her to the window. He tried to get a fireman to give him a rope and help him get her out of the window. But instead of getting the assistance he asked for, he was ordered to get off the ladder. It was Anderson's belief that Mrs. Schiable was not dead at that time.

"The evidence is sufficient to sustain a trial court's finding of negligence. This is not merely a case where the court, in retrospect and using hindsight, had determined that a city might have done things differently in its overall method of fighting a fire. This is a case where the city specifically induced reliance on the skill and authority of its fire department to rescue the decedent, then failed to use due care to carry out its mission, and excluded others from taking action which in all probability would have been successful. It thus placed the decedent in a worse position than when it took complete charge of rescuing her, and became responsible for negligently bringing about her death."

Statutory Immunities and Fire Suppression

Statutory immunities have served to protect fire departments from their expanding exposure to liability. The case of *City and County of San Francisco v. Superior Court* is a good example. The case involved the question of whether or not a city or individual fire fighters working for that city could be held liable for failure to arrive at a fire in a timely manner due to the fire fighters' unauthorized absence from their fire station.

Plaintiffs alleged that "personnel of Engine Company 22, which was located approximately 300 feet from [their] property, participated in an improper social gathering—an abalone dinner—at Engine Company 31 on the evening of the fire, with the result that Company 22's firehouse was left unmanned. The fire was reported by persons who ran directly to Engine Company 22 rather than by calls to Department Headquarters. As a result of Engine Company 22's absence, response to the fire was delayed and extensive damage to the real property of [the plaintiffs] resulted."

Plaintiffs also alleged that a special relationship existed between the neighborhood residents and Engine Company 22, apparently due to their proximity to each other, and further that breach of fire department regulations gave rise to liability.

The court held that the absolute immunity provided by California's statutes absolved the city and the fire fighters of liability under any theory—general negligence, breach of a special duty, or violation of department regulations. As the state statute granting immunity covered not only fire suppression efforts, but also the amount, arrangement, and condition of fire fighting equipment, the court was compelled to find the city and its fire fighters immune from liability.

The San Francisco case is an example of the total protective effect of some immunity statutes. However, California's Code is not typical of the protection afforded in many states. Careful attention is required in determining the scope of application of statutory immunity. As the general rule is liability, with immunity the exception, courts critically focus on the complete details of the fire protection arrangement to make sure the legislature fully intended to grant immunity in a particular instance.

A case exemplary of this critical examination is *Ayres v. Indian Heights Volunteer Fire Department*, involving a suit for negligence against a township trustee and a volunteer fire department. The critical factor in determining if either defendant could be held liable was whether or not they were subject to the protections of the state's immunity statute. The court noted the scope of protection provided by the Indiana Tort Claims Act as protecting "... governmental entities or employees for negligence associated with the performance of a discretionary function. It does not, however, protect independent contractors of governmental entities from liability for negligence associated with the performance of their contractual duties." The court then focused upon the roles and responsibilities of both defendants in the case—the township trustee and the volunteer fire department.

The township trustee had a statutory obligation to provide fire suppression services to the area where the plaintiff resided. State law outlined a number of means for the trustee to provide the required services. An incorporated volunteer fire department was engaged to perform the service. Both were sued for negligence—the trustee in his selection of the means to provide the suppression service, and the fire department in its performance of the service. The court described the background of the incident as follows:

"On the 20th day of January, 1983, at about 10:30 a.m., plaintiffs had a fire in their enclosed ... truck in the driveway of their residence .... The defendant, Indian Heights Volunteer Fire Department, was called by a neighbor of the plaintiffs. Upon the department's arrival at the scene, a neighbor who was extinguishing the fire with his hand extinguisher was told to get out of the way; whereupon the fire fighters sprayed a large fire extinguisher into the rear of the truck with such force that it blew the burning material out of the truck and against the fiberglass door of the plaintiffs' garage, causing it to burn.

"Said fire fighters had a large fire hose, but were unable to get it to work until after setting the garage door afire; then, when they got the hose working, they ignored the request of plaintiffs ... to enter the service entrance and spray from inside the house to keep the fire from entering the garage where the plaintiffs had stored valuable merchandise. Instead, fire fighters sprayed from the outside, blowing the fire from the burning door into the garage and totally destroying the garage and its contents."

The court then focused upon the responsibilities of the township trustee. Recall that he was under a legal obligation to provide fire suppression services to the plaintiffs' property. The law provided that the trustee could discharge his responsibilities by any of the following means:

1. Provide the necessary fire fighting apparatus and equipment, including retention of necessary fire fighters.
2. Contract with a municipality within the township to provide fire suppression services beyond its borders.
3. Cooperate with a municipality in the township in providing fire suppression services (for example, a combined fire department).
4. Cooperate with a municipality in an adjacent township in providing fire suppression services (such as automatic assistance).
5. Contract with a volunteer fire company within the township.

The court found that the selection of the means of providing fire suppression services was discretionary in nature. "It was not a simplistic ministerial act such as the filing of a form or an entry within the record book. The trustee must choose among five delivery systems and without question must exercise considerable judgment and discretion in this choice. The definition of a discretionary duty includes the determination of how an act should be done."

The court did not examine the issue of whether or not the fire department was negligent. However, it did study the matter of whether or not the fire department was protected by the state's immunity statute for performing discretionary acts. The court recognized that decisions on how to fight a particular fire require the exercise of discretionary judgment; specifically, decisions regarding the appropriate fire suppression method or technique to employ for the unique circumstances presented by each fire fighting situation. The decisions normally would be protected activities of a governmental entity's fire department.

However, the court held that the fire department, as an independent contractor for the provision of fire suppression services, was not a governmental entity. Thus it was not protected by the immunity statute. Because all apparatus and equipment was owned by the independent volunteer fire department, and the selection for fire department membership was the exclusive province of the department, the court found the department to be an independent contractor. As such, like any other service provider, it was subject to liability for negligent performance of contracted services. Although the service was performed exclusively for a governmental entity, statutory immunity protection was reserved for government authorities and agencies—not its independent service providers. Had the government agency owned the apparatus and equipment, or controlled the membership selection process, the results could have been different.

Although no blanket rules for fire suppression liability can be formulated, the general parameters to consider include whether or not the department's activities are truly discretionary and require the exercise of professional judgment. Does the fire department have a statutory duty to perform the service for the benefit of all, or only for the protection of a specific group of individuals? Is the suppression service electively undertaken for the benefit of the general public? Has a special relationship been formed, either by contract or by representations to the plaintiff, affirming that he will be protected? Finally, is the fire department immune from liablity under the remaining provisions of sovereign immunity or by statutory immunity?

Fire Drills

Liability for negligent fire fighting efforts is not limited to the emergency setting. As fire departments undertake to increase their live fire training, they can create increased liability exposure. The case of *McCurry v. City of Farmington, New Mexico*, is a good example of this.

During various times from 1977-1979, the City of Farmington and its fire department conducted training exercises. These included the burning of automobiles in a lot located east of and in close proximity to the offices where the plaintiff worked. During the period when the training exercises were conducted, 25 to 35 automobiles were burned. As the court stated:

"The defendants were using machinery and equipment of the Fire Department to train their fire fighters to put out fires they had set themselves. They were using machinery and equipment, which consisted of old automobiles, to train the fire fighters in the use of fire fighting equipment," according to the plaintiff's case. The plaintiff was exposed to the toxic gas and smoke byproducts of these automobile fires, expressed fears as to her resulting health, and sought medical advice.

In January 1979, the plaintiff was advised by her doctor that her symptoms suggested poisoning. After performance of various tests, the plaintiff's doctor determined that she had been in contact with cadmium and other heavy metals in amounts sufficient to cause material health problems. She was admitted to the hospital where medical treatment was undertaken to stabilize her condition, but she still experienced pain in her legs, arms, shoulders, and head; loss of memory; and nervous problems.

The plaintiff's suit contended that the city and fire department were negligent in that they knew—or should have known—of the hazards created by releasing into the air poisonous substances produced by the burning of the automobiles, and that the by-products of such fires would be inhaled by persons in the vicinity. The plaintiff alleged that the fire department also was negligent for not abating the nuisance created by the automobile fires.

The court held that the fire department did not have statutory immunity in this instance. Inasmuch as it employed its own equipment—the junk automobiles—to create the fires, it could be held liable for the negligent maintenance and operation of this equipment.

Emergency Vehicle Operation

Historical Background

Long before the era of streamlined, siren-screaming fire engines, America's cities began to defend themselves in civil actions for injuries arising from the negligent operation of emergency vehicles. People were knocked down even by hand-drawn hose carts. The advent of the steam fire engine brought its all-too-frequent tendency to cause horses to bolt in fright, often running into buggies and pedestrians. Negligent vehicle operation actions soon followed.

Under common law, and in the absence of any statute establishing responsibility, a municipal corporation was held not liable for the negligent operation of fire department vehicles. Until recently, this was the prevailing law throughout the nation, regardless of whether or not the vehicles were being used in the extinguishment of fire. This belief was consistent with the general rule of sovereign immunity for fire suppression activities, as operation of the emergency vehicles was essential to performance of the fire suppression function.

Maintenance and Repair Distinguished

Note that negligence in operation of an emergency vehicle should be distinguished from negligence in its maintenance and repair. Because maintenance and repair work was not considered necessary for the effective performance of the governmental function of fire suppression, as fire fighting can be performed by private parties, immunity protection for negligent maintenance of vehicles and equipment was considered unwarranted. In an early case, *City of Houston v. Shilling*, the Texas Supreme Court explained this distinction and the reasoning behind it in the context of an accident involving a refuse collection vehicle. The accident occurred when the truck's brakes pulled, causing it to move erratically and to collide with Mrs. Shilling's automobile. The refuse truck operator, on multiple occasions including the day prior to the accident, had advised the city's repair shop personnel of the brake problem and requested corrective action. The day following his last report of faulty brakes, the driver was assured that the vehicle had been fully repaired; however, the condition remained and caused the injurious accident.

The city contended that, as it was a governmental function to collect refuse, it was a necessary element of that function to maintain the collection vehicles in good repair; also, it was necessary for the city to maintain a repair shop in order to fulfill the repair obligation.

The court discredited this line of reasoning, stating that such belief would sustain a city's immunity for negligent operation of a gasoline refinery, because fuel is necessary for operation of a refuse vehicle. As a result, the city would be immune from liability involving activities well removed from its true governmental functions. Rather, the court concluded,the line of immunity should be established at the point of demarcation between activities that are truly governmental and those that merely support the performance of the government activity. The case of *Houston v. Shilling* was an early effort to eliminate the harshness of the common-law rule of municipal immunity.

Vehicle Operation—Statutory Modification of Common-Law Rules

To provide certainty regarding the existence/nonexistence of immunity, as well as to soften some of the common-law rules, most states eliminated sovereign immunity in the vehicle operation context in favor of specific statutory protections. Reference to the statutes of a particular state is necessary in any specific case. However, for purposes of illustration, the statutory liability for negligent operation of emergency vehicles under California law, along with references to other states having comparable holdings, will be used here. The laws of other states naturally vary somewhat, but many of the issues presented and analyzed here are illustrative of those in all the states.

The California Legislature has imposed liability on the state and every city or other public corporation owning or operating any motor vehicle. For death or injury to persons or damage to property, the liability applies to damage resulting from negligent or wrongful acts or omissions in the operation of motor vehicles by employees of the corporation, acting within their scope of employment.

Comparable liability is created in other states which have declared public employees to be subject to liability to the same extent as private persons, with their employing jurisdictions mandated to assume the employees' liability obligation under *respondeat superior* principles.

Careful examination of each liability statute is required to determine when the government entity will be liable. For example, the courts have held that in order to recover against a city under California's statute, it must be shown that at the time of the accident the driver of the city vehicle was acting within the scope of his authority. In one case, an off-duty Los Angeles police captain took his wife out to dinner in a police car. He negligently drove into the automobile of a magician, who was injured. The court found the city not liable, as the captain was acting outside the scope of his employment, but the captain remained personally responsible.

In another case, the court found that driving a Los Angeles Fire Commissioner from his home to his office, as ordered by the driver's captain, was within the scope of the driver's authority; hence, municipal liability was properly imposed. The city had contended that the transportation of a city official to and from his home was unauthorized and unlawful. The court did not deny this allegation, but found that the Commissioner had gone directly from his home to inspect a fire station. This had the effect of giving authorization to the entire trip.

Before liability can be imposed under the California statute, and all other comparable state statutes, injury in the case must have been caused by negligent operation of the emergency vehicle. However, what constitutes negligence in the operation of a responding emergency vehicle and of a normally operating private or public vehicle are materially different. Whereas the former is performing a governmental service of potentially immeasurable importance, the latter may be operating for purely private, self-serving purposes. Thus, what does one consider in determining when a city and its emergency vehicle operator should be liable?

A simplified means for potentially finding an individual negligent is to show his violation of a statutue designed to protect the public. "When a statute provides that, under certain circumstances, particular acts shall or shall not be done, that statute may be interpreted as fixing a standard for all members of the community, from which it is negligence to deviate."[3] The adverse impact of this rule should be readily recognized, as emergency response frequently involves violation of statutory rules of the road. Note also that the "violation of a statute raises a presumption of negligence which may be rebutted by proof the violator did what might reasonably be expected of a person of ordinary prudence, acting under similar circumstances, who desired to comply with the law." (*Grant v. Petronella, California*)

To overcome this presumption, states have passed exemption statutes—statutes which exempt responding emergency vehicles from compliance with specified vehicle code provisions, principally those involving speed, direction of travel, and stopping at traffic control devices. The importance of these exemption statutes is their effect of removing the presumption of negligent driving by emergency vehicle operators during emergency response.

Kansas Statutes Annotated, Section 8-1506, is an example of an exemption statute:

a. The driver of an authorized emergency vehicle, when responding to an emergency call or when in the pursuit of an actual or suspected violator of the law, or when responding to but not upon returning from a fire alarm, may exercise the privileges set forth in this section, but subject to the conditions herein stated.
b. The driver of an authorized emergency vehicle may:
    1. Park or stand, irrespective of the provisions of this article;
    2. Proceed past a red or stop signal or stop sign, but only after slowing down as may be necessary for safe operation;
    3. Exceed the maximum speed limits so long as such driver does not endanger life or property;

---

[3] Prosser, *op.cit.*

4. Disregard regulations governing direction of movement or turning in specified direction; and
5. Proceed through toll booths on roads or bridges without stopping for payment of tolls, but only after slowing down as may be necessary for safe operation and the picking up or returning of toll cards.

c. The exemptions herein granted to an authorized emergency vehicle shall apply only when such vehicle is making use of . . . audible . . . and visual signals . . . .

d. The foregoing provisions shall not relieve the driver of an authorized emergency vehicle from the duty to drive with due regard for the safety of all persons, nor shall such provisions protect the driver from the consequences of reckless disregard for the safety of others.

These exemption statutes thus serve to eliminate the presumption of negligence by violating a vehicle operation law. However, it is important to recognize that the exemption statute, while granting the emergency vehicle operator exemption from various travel restrictions, directs that an emergency vehicle driver nevertheless has a duty of safety for the protection of the public. This should never be overlooked or forgotten. The following is a review of various cases that have addressed comparable exemption statute provisions. These cases should enhance understanding of the implication of such statutes.

1. *The driver must be operating an authorized emergency vehicle.*

The meaning of an "authorized emergency vehicle" might be defined by statute. For example, Indiana Code Section 9-4-1-2(d) defines "authorized emergency vehicles" as:

1. Vehicles of the fire department, police vehicles, and ambulances and other emergency vehicles operated by or for hospitals, or health and hospital corporations . . . .
2. Vehicles other than ambulances which are owned by persons, firms, or corporations other than hospitals, and are used in emergency service, may be designated as emergency vehicles if such vehicles are authorized to operate as such by the Department of Highways . . . .
3. Ambulances which are owned by persons, firms, or corporations other than hospitals and which are approved by the Indiana Emergency Medical Services Commission . . . .

Other states' designations of "authorized emergency vehicles" sometimes are more encompassing and can expressly include vehicles used for rescue, ambulance service, fire fighting, lifeguard services, lifesaving, law enforcement, emergency towing and repair, fish and game law enforcement, etc.

Where the designation of an emergency vehicle is not prescribed by statute, the court may apply a "reasonableness standard." "Reasonableness requires us to hold that a 22-foot fire truck, painted bright red, with a beacon light and two flashing red lights, and a siren heard by persons a block-and-a-half or two blocks away, is an emergency vehicle." (*Shawnee Twp. Fire District No. 1 v. Morgan, Kansas*)

Not all vehicles of a fire department would be considered emergency vehicles. For instance, vehicles used by members of a fire prevention bureau, a hydrant inspection detail, or a photography or public relations unit might not be considered authorized for emergency status. On the other hand, the private car of the chief of an organized fire department, equipped with a siren and warning lights and used in responding to emergency fire alarms, would be an authorized emergency vehicle. Reference to a specific state's particular scheme for designating authorized emergency vehicles is required.

2. *Vehicle must be responding to an "emergency call" or alarm of fire.*

The general rule is that there need not be an "emergency in fact"; rather, the driver only needs to reasonably believe that a genuine emergency or fire alarm has occurred.

In one case, an ambulance carrying a man to the hospital for an appendicitis operation collided with the co-defendant's vehicle and injured the patient/plaintiff. The court held that the fact that the operation was not performed for hours after the patient's arrival at the hospital was inadmissible, as this was immaterial to the question of whether or not the ambulance was on an emergency call. The court reasoned that the test for determining if a publicly owned ambulance was an "emergency vehicle" within the statutory exemption from the rules of the road is not whether an "emergency in fact" existed, but rather whether the ambulance was being used to respond to an emergency call. Thus it should be immaterial to the issue of municipal liability whether or not the call subsequently proved to be a false alarm.

Test driving a vehicle, relocating of apparatus from one station to another, or performing in-service inspections generally will not be considered responding to an emergency. Nor will response to a command to "come to my office" be considered an emergency, although the subject to be discussed may be of major personal importance.

Does an emergency call include rescuing a person from a cave-in, removing a cat from a telephone pole, resuscitating a drowning victim or a new-born baby, pumping-out of a cellar, assisting police in apprehension of a fugitive, or protecting a roof with a salvage cover? All these and many other kinds of calls are not uncommon to the modern fire department. However, examination of the particulars of each incident might indicate whether or not there were reasonable grounds to believe that an emergency did exist.

To be classed as an emergency vehicle, must the apparatus be responding to a call which is within the fire department's authority to answer? Is a fire department vehicle, such as a rescue apparatus, an authorized emergency vehicle when it responds to a medical call? Such questions arise if the authority given fire departments by statute or charter extends *only* to extinguishing and preventing dangerous fires—as is the case in many cities. A fire department's activity might be considered unauthorized if it extends to matters unrelated to extinguishing or preventing fires. Courts might be constrained to hold certain activity authorized where no legal authority exists for the department to perform such service. However, in the author's opinion, no doubt should exist that any rescue or first aid activity at the scene of a fire—whether for the benefit of fire fighters or civilians—is clearly within the implied powers of the fire department, as incidental to performance of the fire suppression function.

Sometimes courts have held that a municipally-owned rescue apparatus was not an authorized emergency vehicle, because it was not acting within the scope of the authority extended to the fire department by the city charter. This might occur when an ambulance was responding to a private home for a routine transport, or when a fire department salvage vehicle was responding to a call to place a cover on a leaky roof.

In such cases, the drivers would not be entitled to the statutory exemptions from the rules of the road. Judgment of their conduct would be based upon the "reasonable person" standard, with the understanding that the drivers were without authority to utilize emergency warning devices. Also, these drivers had an obligation to exercise due care to all others without consideration of the nature or purpose for which they were traversing the roads. Under such a holding, the victim of a collision caused by the operation of such a vehicle probably would recover damages from the vehicle operator.

3. *The siren must be sounded.*

Some confusion has arisen over the use of expressions in vehicle codes requiring the operator of an authorized emergency vehicle to "sound an audible signal by siren," or saying "upon the immediate approach of an authorized emergency vehicle giving audible signal by siren." It is the sounding of the audible signal, along with any required display of a visual warning device, that exempts the emergency vehicle operator from certain rules of the road. Performance of both acts also places a concurrent obligation upon all other vehicle operators: to move to the right side of the road, to stop, and to yield the right of way to the emergency vehicle. Thus, in effect, it is the sounding of the siren and display of the visual warning device that serves to begin transfer of the right of way to the emergency vehicle operator.

Failure to sound the siren or display the required warning light means that a vehicle driver cannot make use of the benefits of the exemption statute. In

Grant v. Petronella, a sheriff's vehicle, without using audible or visual warning devices, was engaged in a high-speed response on a multi-lane highway when it struck the plaintiff's automobile. At trial, the sheriff's deputy testified that he did not display a warning light or sound the siren because he believed that these were unsafe procedures while he was engaged in a high-speed response on a highway. The deputy said he feared the response of drivers who, when hearing the rapid approach of the sheriff's vehicle, would quickly change lanes to relinquish the right of way. The court discounted the deputy's reasons for not using vehicle warnings. It held that provisions of the vehicle code section exempting emergency vehicles from rules of the road do not apply when an emergency vehicle has not exhibited a warning by light, siren, or other means.

There is a conflict among jurisdictions regarding whether or not the other vehicle operators must actually hear the siren or see the warning light before the obligation to yield the right of way arises. Most states hold that the giving of the signal by the emergency vehicle operator is his measure of care; if this is done, his duty of care is fulfilled. In a Los Angeles case, the operator of a car in which the plaintiff was a passenger saw a police car approaching an intersection, heard its siren, and assumed the police car could stop in time to avoid a collision. The civilian driver thought "he could beat it across the intersection." The court held that, inasmuch as a warning siren had been sounded, negligence of the police driver could not be a factor in his failure to obey the stop signal or his fast rate of speed. The court said: "Our conclusions from the foregoing are that once an emergency vehicle responding to an emergency call gives the statutory notice of its approach, the employer is not liable for injuries to another unless the operator had made an arbitrary exercise of these privileges." (*Lucas v. Los Angeles, California*)

However, this position is not held by all states. The Delaware case of *Green v. Millsboro Fire Company* involved a collision between a pickup truck operated by plaintiff Willie Green and a fire truck owned by defendant Millsboro Fire Company. The accident took place early in the morning at the intersection of two multi-lane highways. Road conditions were excellent, as it was a bright summer day. Mr. Green slowed down below posted advisory speed before entering the intersection where the collision occurred. He claimed never to have heard or seen the oncoming fire truck. The apparatus driver, with warning lights operating, sounded the siren before entering the intersection but indicated he did not see Mr. Green's vehicle prior to contact.

In refusing to hold Mr. Green negligent as a matter of law, and directing a jury trial on the question, the Delaware Supreme Court noted the plaintiff's contentions that the angle of approach of the two vehicles made it potentially impossible to hear the siren; that the bright daylight made observance of the fire truck's warning lights difficult; and that obstructions along a portion of the highway immediately before the intersection prevented observance of any warning lights. The court held that before a private vehicle driver's obligation to yield the right of way accrues, it must be established that the individual was aware or should have been aware of the emergency vehicle's approach. Although a driver has an obligation to maintain an effective lookout, he must have an opportunity to hear and observe the warnings of an emergency vehicle before his obligation to yield the right of way arises.

Some states require sirens to be an approved type. Other states only give a performance requirement to be met for the siren to be approved—typically that the siren, exhaust whistle, or bell be capable of giving an audible signal loud enough to be heard 500 feet. Failure to use an approved type of siren rarely has prevented designation of an apparatus as an emergency vehicle, provided the vehicle is capable of giving a suitable warning. Courts hold that state legislatures did not intend to deny an emergency vehicle the privileges and immunities granted by statute simply for failure to comply with the formality of obtaining approval for the type of siren used. However, the fire department is responsible for making sure its sirens and other warning equipment meet the performance requirements or implementing regulations of state statutes.

4. *A warning light must be displayed.*

If the warning lights are turned on by the emergency vehicle operator, the fact that the lights are rendered invisible by weather or sunlight conditions, or are not seen because of the other driver's inattention, would be immaterial in a court case. However, as with sirens, some opportunity to see the warning light may be required. Also, the warning light must be of the type and color designated statutorily or must give a "reasonable warning" for a minimum of 500 feet on a clear day.

5. *The operator must drive with due regard for the safety of all persons and must not arbitrarily exercise his privileges.*

The standards of care charged to the driver of an emergency vehicle and to the driver of an ordinary vehicle are not the same. The privileges and immunities granted to the driver of an emergency vehicle are not available to the driver of an ordinary vehicle. Notwithstanding the grant of immunities, all emergency response statutes require the operator of an emergency vehicle to drive with due regard for the safety of all persons. The statutes specify that the immunities do not protect the driver from the consequences of his reckless disregard for the safety of others. (*Shawnee Twp. Fire District No. 1 v. Morgan*, quoted above)

The typical vehicle code thus sets a separate standard of care for the drivers of emergency vehicles, making it lawful for them to disregard stop signals, exceed speed limits, travel in a restricted direction, or stop where others may not. The vehicle code's admonition to emergency vehicle drivers to show due regard for the safety of others on the highway remains an overriding requirement, however. The privileges and immunities granted cannot be exercised where they will present an unreasonable danger to the public.

Due regard for the safety of all persons means that the emergency vehicle driver must, by suitable warning, give others a reasonable opportunity to yield the right of way. Again, this means to give all others an opportunity to escape the path of the emergency vehicle and not endanger anyone else. Where the movement necessary to yield might endanger others, the emergency vehicle operator cannot exercise his privileges. Rather, he must yield for the safety of others.

To summarize, emergency response should be considered as the orderly relinquishment of rights to the emergency vehicle operator, rather than his authorized taking of rights without concern for the consequences.

Arbitrary Exercise of Privileges

The expression "arbitrary exercise of privileges" has caused some confusion. Since emergency vehicles are excluded from the restrictions on speed and right of way, the charge of negligence cannot be based on violation of those restrictions if proper warnings have been given. This is among the privileges granted by the exemption statutes. However, such privileges cannot be utilized arbitrarily; only existence of an emergency warrants their use. An "arbitrary exercise" finding may rest upon the question of whether or not an emergency actually existed. Exercise of these privileges while returning from a fire, or while driving for some private purposes of the operator, might constitute "arbitrary exercise of privileges." (*Lucas v. Los Angeles*, cited earlier)

The view has been advanced that the practical purpose of the phrase "arbitrary exercise of privileges" is to cover cases in which the driver of an emergency vehicle—by virtue of the character of the vehicle, and by use of its siren—maintains unnecessary speed and needlessly drives with abandon of the usual rules of the road. Such is the case when a fire engine is returning from a fire in the absence of any new call or emergency; when the vehicle is being warmed up or tested; or when the chief is being rushed to his office. Add to these situations one where the operator gives the required signal, but sees that another driver has neither heard nor heeded it. Under such circumstances, it no doubt would be considered an arbitrary exercise of privileges to continue toward an inevitable collision.

Pedestrians as well as vehicle operators are under no obligation to anticipate the presence of an emergency vehicle unless they know, or by reason of warning lights and sirens should know, that an emergency vehicle is approaching. This fact, along with an emergency vehicle operator's obligation to drive with "due regard for the safety of all persons," has resulted in many departments requiring drivers to observe all traffic control signals even when they have the right of way. As of this writing, there has been no determination on the question of whether or not an apparatus operator's failure to comply with department regulations, all traffic signals, or speed limits could serve as a basis for liability.[4]

Exemplary Cases Involving the Law of Emergency Vehicle Operation

California—In this case involving the Los Angeles Fire Department, Engine 22 and Engine 14 collided at Jefferson and San Pedro streets during a response to a street box. Both vehicles had their red lights and sirens on and both drivers were familiar with their prescribed response routes. Engine 14 was going west on Jefferson and Engine 22 was going north on San Pedro. Each knew the other would be coming. The driver of Engine 22 had seen Engine 14 pass in front of him many times (at least six) on previous runs.

LAFD rules required all rigs to slow down at intersections, and if a "Stop" light were showing, to stop if necessary; in any case, extreme caution must be used. In this instance, the stop light was against Engine 22; its Acting Captain was the first to notice Engine 14 coming from the right; and its driver, when asked if he ever looked to his right in the last 100 feet of travel, said "I can't say that I did or didn't." In a prior deposition, the driver had stated "If the rig had been there, I would have seen it if I had looked." He also testified that he had not paid any attention to the signal and that "We carry our own signal with us."

Engine 14 hit the right rear of Engine 22, causing it to collide with the Torres auto which had pulled over and parked in obedience to the engine's warning devices. The Torres auto in turn struck the Baskett auto, which also had stopped for the same reason. The collision resulted in the death of Mrs. Torres, injuries to Mr. Torres and his daughter, and injuries to both Mr. and Mrs. Baskett. The jury rendered judgments for plaintiffs. It held that although statutory exemptions to some rules of the road existed, the court found no blanket exemption from all the rules that govern the operation of motor vehicles. It held that the duty to drive with due care for the safety of all persons had been breached. (*Torres v. Los Angeles*)

Tennessee—The court of appeals affirmed judgment for the plaintiff, a motorcycle operator, against the fire department. Evidence showed that, although displaying the required warning devices, a city fire truck failed to slow down before entering a hazardous intersection against a red light. This failure to reduce speed was the proximate cause of the truck's collision with the motorcycle. (*Green v. City of Knoxville*)

Ohio—A case arose out of a collision between an automobile driven by the plaintiff and a truck driven by the defendant volunteer fire fighter. On the evening of the accident, the volunteer was at home when he received an emergency call that a house was on fire in the village. He was on his way to

---

[4] See General Annotations, "Liability for Personal Injury or Damage from Operation of Fire Department Vehicle," 82 *American Law Reports* 2d 312.

the fire station in his pick-up truck with red lights and siren in operation when the plaintiff drove his car into the truck's path. The two vehicles collided and both parties were injured. When the plaintiff sued to recover for his injuries, the trial court held the volunteer immune from liability under Ohio Revised Code 701.02.

Ohio statute 701.02 provides in relevant part: "The defense that the officer, agent, or servant of the municipal corporation was engaged in performing a governmental function, shall be a full defense as to the negligence of . . . members of the fire department while engaged in duty at a fire, or while proceeding toward a place where a fire is in progress or is believed to be in progress, or in answering any other emergency alarm. Firemen shall not be personally liable for damages for injury or loss to persons or property and for death caused while engaged in the operation of a motor vehicle in the performance of a governmental function."

When asked whether the immunity was applicable to volunteer fire fighters while operating their personal vehicles, the Ohio Supreme Court held the plain language of the statute was controlling and that volunteers were fire fighters within the meaning of the immunity statute and thus protected by the statute. (*Dougherty v. Torrence*)

Wisconsin—The defendant fire fighter, although privileged to violate rules of the road in response to an emergency, was found liable when such activities were executed while in a state of intoxication. The court held: "No statute authorizes a fire fighter to drive while under the influence of an intoxicant in the performance of his or her duty. No part of the duties of a fire fighter apparently authorizes him or her to drive while intoxicated. No case suggests that fire fighters are apparently authorized to drive while intoxicated. Common sense suggests the opposite is true in a situation where judgment and skill are required." (*State of Wisconsin v. Schoenheide*)

New York—A plaintiff, while sitting in her legally parked vehicle, was struck by one of the defendant's fire trucks as it attempted to negotiate a turn. The fire fighter directing the truck as it turned testified that he did not see the parked car but admitted that he should have seen it. The city was held liable for negligent operation of an emergency vehicle. No privileges or immunities were found applicable to nonemergency operation of a fire truck. (*Thompson v. City of New York*)

Illinois—A jury returned a verdict for the estate of the plaintiff motorcycle operator who was struck and killed by the defendant city's fire truck. The apparatus entered an intersection, reportedly against a red light, and without sounding a siren, operating warning lights, or reducing speed. Total award exceeded $1.2 million. (*Almanza v. City of Chicago*)

Fireboats

Many fire departments today have fireboats in service, so it is important to note that the old common-law rule of nonliability for fire department functions does not extend to this branch of the fire service. Under maritime law, a city is liable for the negligence of its fire fighters in operating its fireboats. At an early date, the United States Supreme Court ruled on the case of *Workman v. New York*. The "New Yorker," a steam fireboat, collided with the barkentine "Linda Park" while the fireboat was running into a slip to extinguish a fire in a warehouse. The court declared that "under the general maritime law, where the relation of master and servant exists, an owner of a vessel committing a maritime tort is responsible under the rule of *respondeat superior*."

Sending Fire Trucks Outside the City Limits

Many cities make a practice of sending fire fighters and apparatus outside the city boundaries in order to assist neighboring city, county, or Federal government fire departments. Although mutual assistance generally is rendered in a spirit of good will, it also can be performed under terms of a contract or other fee arrangement. In such cases, fee payments may be applicable based upon the number of fire fighters sent, the fire trucks or other equipment used, and the time spent at the fire.

These mutual aid contracts are not considered binding upon the fire departments unless the department officials negotiating such agreements have the authority to make such contracts—authority granted by state law, city charter, or municipal ordinance authorized by state law or state constitution provision. The general rule is that a municipal corporation can exercise only those powers conferred upon it by charter or general law of the state. As the powers of a city or other incorporated municipality are limited to application within the municipality's jurisdiction, special powers must be granted for the entity to perform any services outside its boundaries.

In *City of Pueblo v. Flanders*, a court addressed the question of extra-territorial response authority. The court held that the authority to declare policy concerning use of the city's fire equipment and fire fighters for extinguishment of fires outside the city limits rests with the city council; until the council acts, the decision to or not to fight extra-territorial fires is at the discretion of the city's responsible officials. The majority opinion inferred that such outside assistance was justified from a humanitarian standpoint, and ruled, in a taxpayer's suit, that the city could not be enjoined from rendering such service.

The court stated: "The distinction, we think, is fundamental, and in cases involving actions of municipal authorities outside the city limits, it is important to distinguish which question may be involved: the power to enforce the authority of the municipality against property or people outside its borders; or the right of municipal officers to act outside the city limits where no enforcement of authority is involved. The former right, if existing at all, is absolute [created by statute]; the latter is discretionary and dependent on purpose and result. The former. . .is dependent on legislative delegation of authority; the latter is dependent on whether or not the use of the city's officers or facilities or funds is, or is not, for the benefit of the municipality....

"As to the second ground, we cannot agree with the unsupported declaration of the trial court that the welfare and public interest of the municipality and the taxpayers therein are neither promoted or protected by permitting the city's fire department to accept calls for the extinguishment of fire outside the municipality's corporate limits, even where public buildings are not involved. In many cases prompt action in extinguishing a small fire outside the city limits may prevent its increase and spread across the city line with disastrous results to the city and its taxpayers. The destruction by fire of a factory located outside the city limits may deprive resident taxpayers of means of livelihood and they and the city suffer loss thereby.

"Mutual assistance by neighboring cities may well work to the advantage of each. Perhaps in some cases the very good will acquired by assistance outside the city in an emergency may be of value to the city and its taxpayers...."

Must a city go outside its boundaries to fight a fire? Unless the fire department is bound by an ordinance or statutory requirement or a contractual agreement to respond, the rule is no. In fact, some states have statutes specifically permitting or prohibiting such activities. The more important question is, does the city have the power to send its apparatus outside its limits, and if so, what legal implications may arise as a result of undertaking such a practice?

In *Canade v. Town of Blue Grass*, property owners brought suit against the town and the captain of the responding fire crew of the volunteer fire department. The charge was that they left the scene of a fire on the plaintiff's property without attempting to extinguish the blaze. The court held that since the town would be immune from liability for negligence in fighting a fire within its limits, likewise it could not be held liable for failure to extinguish a fire outside its limits. The plaintiff was thus precluded from recovery despite the fact that the volunteer fire department responded outside the city's limits, determined that the plaintiff was not a subscriber to its services, and returned to quarters without attempting to perform any fire extinguishment. The court majority held that the city did not forfeit its immunity upon leaving the city limits in this case, even though it performed contractual fire suppression services to various other parties outside city limits. No implied contract to extinguish the fire could be formed based upon the existence of the fire emergency, the fire department dispatcher's assurance that the fire department would respond, and the fire department's presence at the site. The fact that the plaintiff detrimentally relied upon the department for fire suppression efforts, in lieu of calling another fire department, did not overcome the finding of immunity.

A highly critical dissenting opinion stated that, because the plaintiff and defendant were at an impasse over fees to be charged for the fire suppression service, the defendant fire department was more interested in "soaking the plaintiff than in soaking the subsequent fire." The court could have found liability on an implied contract or special relationship theory. It also could have applied a fundamental theory of law: "One who undertakes, gratuitously or for consideration, to render services to another which he should recognize as necessary for the protection of the other's person or things, is subject to liability to the other for physical harm resulting from his failure to exercise reasonable care to perform his undertaking, if a) his failure to exercise such care increases the risk of such harm, or b) the harm is suffered because of the other's reliance upon the understanding."[5]

Although the court sustained the general rule of nonliability, the numerous dissents indicate that caution must be exercised in the rendition of express or implied contract fire suppression services. As discussed earlier in *Ayres v. Indian Heights Volunteer Fire Department*, where the fire department specifically contracts to perform the fire suppression service, it can be held accountable for negligence in the rendition of that service.

In *Tuthill v. Rochester*, the city was held liable to persons whose properties were destroyed by fire. Here the city had a contract with a suburban fire district to provide it with fire suppression assistance. The city, however, failed to honor its contractual obligation to the district. Note the distinction of this case from *Canade*, above, where no contract existed and none could be implied. In the *Rochester* case, a contract with the district did exist, and was enforceable against the city based upon an intended beneficiary theory.

Regarding emergency response, courts have held that in sending fire apparatus outside the city under a fee arrangement, the fire department's activity changes from a governmental to a proprietary one. This renders the municipality liable for damage sustained by persons injured negligently by its fire trucks on such a call. It is immaterial that in a particular accident the city received no remuneration despite its general practice of charging for such services.

Others have noted that, in considering the question of liability, where a statute imposes municipal liability for the negligent operation of emergency vehicles, it should be immaterial whether the negligent act is committed within or outside the city limits. The public policy which brought about adoption of such liability statutes should lead to such a conclusion.

Although the subject of workers' compensation insurance is considered in a later chapter, it is relevant at this point to note that some states have adopted statutes which provide that all fire fighters, including volunteers, are eligible for such compensation for injuries incurred while fighting fires outside their jurisdiction. Under the applicable statutes of the State of New

---

[5] *Restatement (Second) of Torts*, Section 323 (1965)

York, for example, it was held that a volunteer fire fighter who was summoned by his own fire district to aid another district, and who was injured in the line of duty, was entitled to workers' compensation paid for by the district that was being aided. (*Gilewski v. Mastic Beach Fire District No. 1, New York*)

Mutual Aid

As stated earlier, the general rule is that a fire department may not operate outside its jurisdiction unless expressly authorized to do so by statute. Several cases have held that, although it is acceptable to obtain fire fighting services from outside sources, a city has no authority to enter into an agreement to help another jurisdiction unless the law grants such authority.

Many states have laws which permit governmental entities to enter into mutual aid agreements to help each other in case of "great peril, extreme emergency, or public calamity"—the typical mutual aid agreement. However, for provision of initial alarm response across jurisdictional boundaries, specific statutory authority is required. Such authority can permit fire departments to assist each other on a day-to-day basis, and to cover-in for an adjacent community whose apparatus is out on a fire or is rendering aid to other neighbors.

Mutual aid agreements can be informal understandings or formal contracts. The latter is preferred and should include an outline of the means for activating the mutual aid arrangement, the chain of command in a multiple jurisdiction fire situation, the reimbursement arrangement for expended materials, and indemnification or hold-harmless agreements in the event of personal injury or property loss.

Problems in Fire Extinguishment

Emergency Dispatch Services

An integral element of the fire department is the emergency dispatch function. As this involves activities both discretionary (such as the number and type of emergency vehicles to dispatch) and ministerial (such as alerting units and notifying them of the emergency's location), courts generally have held dispatch activities to be immune from liability. In the absence of a special relationship between the victim and the fire department, no duty to assure timely response of emergency vehicles will exist. In *Helman v. County of Warren*, the plaintiff's shop was destroyed despite two calls to the newly-operational 911 emergency command center. Neither call was successful in making contact with any dispatch operator. A third party subsequently called 911 to report the fire and was told that the responsible fire department would be notified immediately. It was not until a passing motorist stopped a patrolling deputy sheriff that the fire department actually was notified. The result was that the first-due unit did not arrive on the scene until more than 50 minutes had elapsed since the plaintiff's first notification attempt. Meanwhile, the building was totally destroyed. In holding the county not liable, the court of appeals stated that there was no statutory or general obligation of government to assure an emergency response, and that the "absence of any direct contact between agents of the municipality and the injured party upon which a special duty could be predicated" precluded the finding of any liability. Thus, without the direct contact, the plaintiff did not personally, detrimentally rely upon assurances of the 911 operator that an emergency response was being undertaken.

Again, in *Trezzi v. City of Detroit*, the court followed the no liability rule. However, a vigorous dissent stated that 911 is not essential to the effective performance of a government emergency service. As dispatch services could be provided independently by the respective police, fire, and ambulance services, 911 simply acts as an answering and message forwarding center—truly ministerial functions. Thus the 911 service should not be protected under the government's discretionary immunity provisions. The plaintiff argued that, had the 911 service been provided by a private contractor, that contractor undoubtedly would have been found liable.

However, where direct contact was made between the plaintiff and the defendant emergency dispatch agency, the situation was different. Here, an affirmative representation was made that emergency personnel were responding. This resulted in the plaintiff's reasonable, yet detrimental, reliance upon that dispatcher's representation. If there were, in fact, no response, the emergency dispatch agency could be held liable. In *Delong v. Erie County*, the estate of the deceased plaintiff sued the county for negligent operation of a 911 emergency notification system. The complaint alleged that 911 operators negligently recorded the incorrect address of the emergency, and failed to follow department procedures when they did not repeat to the reporting party the name of the caller, address of the emergency, and basic information on the emergency . In addition, when responding police officers could not find the location, dispatch operators failed to perform required followup procedures. Justice Stewart S. Hancock, Jr., in rejecting the county's contention of no liability, explained "that it was not the establishment of the 911 system standing alone which created a special duty, but rather the holding out of the 911 number as one to be called to secure immediate assistance which gave rise to a special duty owed to any caller. This holding out induced [the decedent] to rely on the telephone service, and the response 'Okay, right away' to her plea for emergency help caused her to rely further .... The fact that after having called 911, [the decedent] remained defenseless in her home [in response to the dispatcher's request to 'stay where you are,' only to be attacked by a burglar], instead of running out the front door to her neighbor's home, indisputably showed that the conduct of the municipality increased her risk of danger." Liability for such failure to exercise reasonable care could justifiably follow.[6]

Failure to Extinguish Fires

When considering the issue of fire department liability for failure to extinguish fires, it is important to determine the principal factor responsible for the failure: specifically, 1. whether the failure was due to the inability of available equipment to handle the emergency; 2. negligence in suppression procedures; or 3. a complete failure to respond. The distinction can be critical, for courts give great deference to communities in determining the size of the fire departments they can support and to fire chiefs over issues of apparatus deployment and appropriate fire suppression procedures, especially under exigent circumstances. (*City of Hammond v. Cataldi*)

Failure to respond, when required, presents another problem. For example, *City of Daytona Beach v. Palmer* found that "to hold a city liable for the negligent decisions of its fire fighters would require a judge or jury to second-guess fire fighters in making these decisions and would place the judicial branch in a supervisory role over basic executive branch, public protection functions in violation of the separation of powers doctrine.

"Decisions as how to fight a particular fire require that judgments be made regarding appropriate methods and techniques for the unique situation presented by that fire. Although it is possible that ministerial actions may be taken in the course of fire fighting, the plaintiffs' allegations do not indicate that the actions they complain of were not discretionary. Therefore, the city is immune from liability as a matter of law . . . ." (*Chandler Supply v. City of Boise*)

"The decisions of how to properly fight a particular fire, how to rescue victims in a fire, or what and how much equipment to send to a fire, are discretionary judgmental decisions which are inherent in this public safety function of fire protection. A substantial majority of jurisdictions that have addressed the issue of governmental liability for asserted negligent conduct in responding to and fighting fires have reached this same conclusion."

Not all state supreme court justices concur on this rule. The dissenting opinion of Justice Raymond Ehrlich in *City of Daytona Beach v. Palmer*, which represented the opinion of the three dissenting justices of the seven-member Florida Supreme Court, is exemplary. It noted: "When that discretion affects government planning, *Commercial Carrier* has ruled it immune from suit. Thus, had the city sent no fire fighters or too few to the scene, that decision would have been immunized as a planning level, strategic allocation-of-resources decision. Once fire fighters were dispatched to the scene, however, all policy decisions had been made. The fire fighters had no govern-

---

[6] See *Duquesne Law Review*, Vol. 22, 1983, pp. 229, 304.

mental discretion to refuse to fight the fire, to 'fight' it with flammable substances, or to deviate in any way from the reasonable standard of care in fighting that fire. The complaint alleges that the fire fighters did deviate from the accepted standard of care in fire fighting practice. That complaint states a cause of action which should not be barred by the majority's sweeping expansion of sovereign immunity."

Thus, as the dissenting opinion shows, the courts are undertaking increased critical analysis of municipal fire suppression operations. Although the courts generally would continue governmental immunity for the discretionary determination of the size and level of public fire protection services, many justices would eliminate immunity for the actual performance of suppression services. In addition, courts perceive limited difficulty in examining questions of failure to properly overhaul to prevent rekindling, failure to respond in timely fashion, interference with fire suppression operations, injury to fire scene bystanders, or failure to respond at all. It is important to note the increasing critical examination of fire suppression operations being undertaken by the courts.

Examples of Fire Suppression Liability Cases

Williams v. City of Tuscumbia—The fire department's delayed response occurred because the fire truck driver went home sick before a replacement driver arrived. The court held that once the city made the decision to maintain a fire department, it owed a duty to all citizens to act in a skillful manner. Failure to maintain a back-up driver constituted unskillfulness.[7]

Steitz v. Beacon—In this case decided by the New York State Court of Appeals, an action was brought against the City of Beacon to recover damages suffered as the result of a fire. The complaint alleged that the plaintiff's property was destroyed because of carelessness and negligence by the city in failing to create and maintain a properly equipped fire department for the benefit of the plaintiff's property. The court held that the responsibility of the city as far as fire suppression services were concerned—based upon the general grant of power to the city—was to protect the public health and safety and to provide for the general welfare of the community. The court also held that, although there was indeed a public duty to maintain a fire department, there was no duty to individual residents nor any suggestion that the people of the city could recover for fire damages to their property due to any failure to maintain a fire department capable of handling all fire emergencies.

Utica Mutual Insurance v. Gaithersburg-Washington Grove Fire Department—Maryland's Court of Special Appeals held that a volunteer fire department could be liable for negligence in failing to control and extinguish an

---

[7] See Antieau, Chester J., *Municipal Corporation Law*, Lawyers Co-Operative Pub. Co., 1983, Vol. 1A, Section 11.70.

outdoor brush and trash fire adjacent to a warehouse. The fire in question rekindled or ignited a second fire, resulting in destruction of the warehouse. The court stated that "the jury could have believed that the negligent failure of the firemen to extinguish the fire on the outside of the . . . building properly was a direct cause of the second fire." However, where there is no evidence that the subsequent fire was caused by the original one, or that the fire department failed to follow proper fire suppression procedures, courts will hold the fire department not liable as a matter of law, and prohibit a jury's consideration of any negligence question. (*Denham v. Sears*)

In the absence of statutory immunity, a basis for fire department liability could be: a fire department's complete failure to respond when the severity of the emergency was known; the specific location and identity of the victim; and when the response of the fire department is assured. There is no uniform rule on this point among the various states. However, an increasing number of states are following the rule as outlined in *Fairbanks v. Schiable*, quoted above, to find that where a special relationship exists between the fire department and the victim, due to reasonable yet detrimental reliance by the victim, the fire department will be held responsible to the victim for any failure to exercise reasonable efforts to control the emergency. As a minimum, such efforts would include calling for mutual assistance when the department is detained at another emergency.

Significant facts that will help establish the existence of a special relationship include: the seriousness of the emergency is recognized by the fire department; the nature of the emergency involved is communicated; the specific location is identified; the department fails to communicate that response will be delayed due to a concurrent emergency; or the department fails to request mutual aid from the nearest available fire department.[8]

Where a fire department is bound by contract to respond and render fire suppression services, the department can be found liable for failure to respond or to exercise due care in its fire suppression efforts. Where the activities of the contract fire department make the fire conditions worse, the department can be held liable. In *Barnum v. Rural Fire Protection Company*, a private, for-profit corporate fire department responded to the plaintiff's store fire. The location was beyond the franchise response area of the private fire department and the plaintiff was not an independent subscriber to the fire suppression service. However, the department had periodically responded to fires where it had no contractual duty, charging 17 times the annual subscription fee for such service. In addressing the issue of potential liability for negligent fire suppression by this for-profit fire suppression company, the court endorsed the *Restatement (Second) of Torts* rule on the sub-

---

[8] See *City of Fairbanks v. Schiable; City of Kotzebue v. McLean; Chambers-Castanes v. King County; Matlock v. New Hyde Park Fire District.*
See also Antieau, *op. cit.*, Section 11.67, pp. 11-119.

ject: Section 323. Negligent Performance of Undertaking to Render Services. "One who undertakes, gratuitously or for consideration, to render services to another which he should recognize as necessary for the protection of the other's person or things, is subject to liability to the other for physical harm resulting from his failure to exercise reasonable care to perform his undertaking, if a) his failure to exercise such care increases the risk of such harm, or b) the harm is suffered because of the other's reliance upon the undertaking.

The court held that, inasmuch as the fire department did not increase the risk of harm by its response—because the fire, unchecked, would have extended much faster and possibly further—there was no basis of liability for the department's failure to exercise due care in fire suppression. In addition, as the private fire department was the only department available, no harm could be found in the plaintiff's detrimental reliance upon its performance. Had another fire department been available, the court's conclusion undoubtedly would have been different. In general, courts give wide latitude to departments for the strategic and tactical decisions related to their fire suppression efforts. On the other hand, courts have little difficulty in finding liability where a department breaches its contractual duty to respond. (*Ayres v. Indian Heights Volunteer Fire Department and Barnum v. Rural Fire Protection Company*)

In *Matlock v. New Hyde Park Fire District*, three fire districts became involved in a dispute over which had the "right" to extinguish the fire. While the dispute was in progress, fire destroyed the plaintiff's property. The arguing districts could be held liable for the loss.

Can the failure to extinguish a fire be considered a deprivation of a person's civil rights and thus subject the municipality to liability under 42 United States Code 1983? In *Jackson v. Byrne*, Chicago's fire department and fire fighters' union were sued for the deaths of plaintiffs resulting from a house fire that occurred during a fire fighters' strike. Non-striking fire fighters and supervisory personnel were deployed into task forces to selected fire stations. The balance of the stations were closed and guarded by police officers. All striking fire fighters were prevented by police officers from entering their fire stations. When fire broke out in the immediate vicinity of a closed fire station, striking fire fighters attempted to obtain rescue equipment from their station, but the police officers prohibited their entry.

In holding the city not liable, the court stated: "Although there were deaths in this case, the state (city and municipal fire department) did not, within the meaning of the Fourteenth Amendment, deprive plaintiff's decedents of life. The fire killed Santana and Tommie Jackson, government officials did not. Our analysis would no doubt be different if government officials set the fire or placed forces in motion which ignited the fire that claimed the lives of the Jackson children .... The Constitution creates no positive entitlement to fire protection. Similarly, when Mayor [Jane M. B.] Byrne dismissed the striking fire fighters, she did not cause the city to default on an obligation recognized by the Constitution."

In addition, "the Constitution does not require a municipality to provide potential rescuers with even the most elementary fire fighting equipment. The fact that the city had fire fighting equipment nearby and still refused rescuers access to it does not change our analysis. Nothing in the Constitution requires governmental units to allow potential rescuers access to equipment it has already acquired. The Constitution only requires that if the city denies its Good Samaritans access to its equipment, it does so in a nondiscriminatory fashion."

The court noted, however, that a deprivation of civil rights could occur through the city's efforts to deploy the nonstriking fire fighters and apparatus. "The case would be a different one if the selection of firehouses to close during the pendency of the strike had been done on a discriminatory basis. Even a random closing of firehouses, the pattern which the city apparently followed, would have to conform to the dictates of minimum rationality."

Failure to Provide Sufficient Water for Fire Protection

Does a property owner have a right to recover for a loss by fire due to an inadequate water supply? The answer to this question depends not only upon the jurisdiction, but upon whether the water is: 1. provided under contract for the specific purpose of fire protection, 2. supplied by a municipally owned or controlled water company, or 3. provided by a private entity. The general rule is outlined in *Steitz v. Beacon*, quoted above, where the complaint alleged that the destruction of the plaintiff's property was due not only to the negligence of the city in keeping a fire department of deficient size, but also negligence for failing to maintain adequate water pressure for protection of the plaintiff's property. The court held that the city was not liable because failure to furnish an adequate supply of water was at most the denial of a governmental benefit and was not the commission of a wrong. "As there was no agreement in this case to put out the fire or make good the loss, . . . liability is predicated solely upon the . . . provisions of the city's charter defining its powers of government." There was indeed a public duty to maintain a fire department, but that was all, and there was no suggestion that for any omission in keeping hydrants, valves or pipes in repair, the people of the city could recover fire damages to their property. (*Steitz v. City of Beacon; Shockey v. City of Oklahoma City; Cross v. City of Kansas City*)

The case of *Nashville Trust Co. v. Nashville* is often quoted as outlining the general rule and its distinctions. It held: "A municipality operating waterworks pursues that activity in a dual capacity. So far as it undertakes to sell water for private consumption the city engages in commercial venture, functions as any other business corporation, and is liable for the negligence of its employees. However, insofar as the municipality undertakes to supply water to extinguish fires, it acts in a governmental capacity and cannot be held liable for negligence on the part of its employees."

A minority of jurisdictions would hold cities liable for failure to provide adequate water supplies for fire suppression. In *Veach v. Phoenix*, the Arizona Supreme Court expressed this expanding rule: "A municipality has no absolute duty to provide water for fire protection purposes to its inhabitants. However, when a city assumes the responsibility of furnishing fire protection, then it has the duty of giving each person or property owner such reasonable protection as others within a similar area within the municipality are accorded under the like circumstances."

In *DeFrancesco v. Western Pennsylvania Water* the court held that an exculpatory clause in the water company's tariff, limiting the company's liability, was void as being contrary to public policy. The court noted: "Certainly, appellant would have little incentive to ensure an adequate water supply if it were exculpated from all liability for damages resulting from an inadequate water supply. In addition, since appellant owns and has exclusive responsibility for the maintenance and repair of its water mains and hydrants, it is in the best position to reduce the risk of inadequate water supply."

An interesting aspect of this issue arises where the public water supplies are inadequate to support installed fire sprinkler systems. In one instance, a fire that broke out in a property radiated to the plaintiff's property, causing the fusing of interior sprinkler heads. Although the sprinkler system apparently was installed based upon then-available water supplies, subsequent water supplies evidently were reduced materially and became incapable of fully supporting the system. In applying the rule in the Nashville Trust case, the court held that provision of automatic sprinkler system water was comparable to providing water for fire extinguishment purposes and thus the city could not be held liable for the system's inability to contain the fire. (*Gans Tire Sales v. City of Chelsea*)

A distinction must be made between provision of adequate fire fighting water supplies and operation and maintenance of a water distribution system. Although provision of fire hydrants and a water distribution system capable of controlling major fires is a discretionary matter, a water authority must exercise reasonable care in maintaining such equipment after it has been installed. (*Harris v. Board of Water and Sewer Commissioners of Mobile*)

The courts in a majority of states, as well as the Supreme Court of the United States, hold that a private property owner cannot recover against a private water company, whose contract is with a city, for loss by fire on account of the company's negligence. Lack of privity of contract between the property owner and the company precludes recovery.[9] The property owner

[9] See cases collected in 3 *George Washington Law Review* 270 (1935), "Municipal Corporations: Right of Property Owner to Sue for Loss Because of Inadequate Water Supply."

See also 163 *American Law Reports* 348, "Liability of Municipality for Fire Loss Due to its Failure to Provide or Maintain Adequate Water Supply or Pressure."

was not intended to be a third party beneficiary of the benefits of the contract. Rather, any benefits to the plaintiff were only incidental to the water company's performance of its contract with the city. (*Ukiah v. Ukiah Water Co.; Niehaus Bros. v. Contra Costa Water Co.; Heieck and Moran v. City of Modesto*)

Other jurisdictions holding contrary views would permit a property owner to sue as an incidental beneficiary under the water supply contract. As the court stated in *Paducah Lumber v. Paducah Water Supply*: "A party for whose benefit a contract is evidently made may sue thereon in his own name, though the engagement be not directly to or with him . . . [and allowed a recovery against the defaulting water company which had a contract with the city]."

Failure to Inspect for Closed Valves

Because a valve was closed in a water main which supplied a fire hydrant in the water system operated by the City of Grand Coulee, Washington, fire destroyed all the plaintiff's buildings including a laundry, restaurant, and tavern. The Court of Appeals of Washington, in granting judgment for the plaintiff, held that a city which maintains a water system to which fire hydrants are attached has a duty to regularly inspect that system to ensure an adequate water supply to those hydrants. Failure to inspect is a breach of duty as a matter of law.[10]

A city and a civil engineering firm that tests water distribution systems for adequacy and water leakage may be liable for closure of water system valves, for testing purposes, without developing an emergency plan for their reopening in the event of fire. In *Printed Terry Finishing Co., Inc. v. City of Lebanon*, the city engaged the co-defendant engineering firm to supervise a testing program. Its purpose was to determine the location of water leaks within the city's water distribution system. A benefit of the undertaking was reported to be "increased fire protection in the areas adjacent to defective valves which would be either opened or repaired during the survey." Testing required the programmed closure of system key valves, by city employees, to permit hydraulic testing of individual districts of the water distribution system.

The day prior to the fire's outbreak, the city closed various water valves in the district adjacent to the plaintiff's property, which resulted in numerous complaints of low water pressure. The following day, the valves were again closed, resulting in low water pressure supporting the sprinkler system protecting the plaintiff's structure. When a fire occurred, it resulted in major destruction due to inadequate water supplies to support the sprinkler system or fire fighting efforts.

---

[10] Morse, H. Newcomb, "Legal Insight: Dry Hydrant Costs City," *Fire Command*, Vol. 41, No. 5, May 1974, p. 28

Although the general rule is no liability on the part of cities for failure to provide adequate amounts of water for fire protection, the court distinguished this case from the general rule. Liability in this action was premised first upon the city's failure to use reasonable care to adequately maintain the existing water system. The second basis of liability was failure to establish an emergency plan to reopen closed valves in the event of an emergency. As the outbreak of fire is always a possibility, and the closure of water valves materially increases fire loss potential, the city could be held liable for knowingly failing to maintain the water distribution system in full working order.

What losses are recoverable from a city as a result of its negligent closure of water supply system valves? Losses recoverable will be limited to those that were foreseeable at the time of the negligent act. In *Jolly v. Insurance Co. of North America*, a fire that broke out at the plaintiff's property was almost contained by fire fighters, without much resulting damage, when their apparatus tanks ran out of water. Fire fighters attempted to obtain water from the nearest fire hydrants but were unsuccessful. Cora Jolly watched her house being destroyed by fire while the fire fighters stood by, apparently without water needed to contain the redeveloping blaze. She became agitated, which aggravated a preexisting medical condition, then she collapsed and died at the scene. The court limited damage recovery to the Jolly property damages, holding that only such damages were foreseeable as resulting from the failure of the water authority to maintain the fire hydrant adjacent to the property.

Fire Hydrants

Closely related to the question of failure to furnish an adequate water supply are the problems arising from municipal ownership of fire hydrants. In some states, where an injury was caused not by a fire but by a city's negligence in the installation, maintenance, or use of its fire hydrants, the municipality has been held liable even though such hydrants are used for governmental purposes.

Where a loss was sustained as a result of a fire department using a hydrant in extinguishing fires—as, for example, where water ran from a hydrant during the course of fire pumping operations—the city was not liable. This was true even though such hydrants were used on other occasions for other municipal purposes, such as flushing streets, filling city equipment, or watering trees. In the absence of a statutory obligation to protect individuals during the course of fire suppression operations, a city is not liable for the negligence of its fire fighters when they are performing the governmental function of fire suppression.[11]

---

[11] See 113 *American Law Reports* 661, Annotation, "Liability of Municipality for Damage to Person or Property Due to Hydrant."

Municipalities have been held liable in other cases where reasonable care in making inspections would have revealed defects in hydrants or the pipes connecting them to the water mains. Examples would be where the lateral pipe breaks and causes flooding of a nearby cellar; or where the water percolates through the soil, causing a building to settle and crack. This is true even where a pipe which flooded a cellar burst because of added pressure thrown on the system through pumping operations during a fire. Such extra pressure should have been anticipated.

A municipality also is liable when pedestrians injure themselves on unguarded fire hydrants that obstruct the sidewalk. However, a city is not liable for the unknowing, unauthorized use of a fire hydrant which results in pedestrian injury. Where children opened a fire hydrant and directed the stream, by using tin cans, onto the stoop of a plaintiff's home, the city was held not liable for her subsequent fall. (*Irizarry v. City of New York*)

Fire Hydrant Rental and Related Water Charges

Even where a fire department does most of the inspection, testing, and painting of fire hydrants, the department often is required to include in its annual budget enough money to pay the monthly rental fee charged by the water company, whether publicly or privately owned. Frequently, the obligation to pay must be borne by the governmental entity that mandated the provision of the fire hydrants in the first place. In *Sheridan Acres Water Co. v. Douglas County*, a water company endeavored to collect hydrant connection fees from the county. As the water company was required, by state statute, to provide a reasonably adequate supply of water at reasonable pressures for fire protection at all locations where it provided domestic service, it believed it was entitled to a fire hydrant standby charge. The Nevada Supreme Court disagreed, holding that although the water company was obligated to provide adequate fire fighting water supplies, only the entity requiring provision of the fire hydrants could be assessed the hydrant fee. The court record does not so indicate, but it reasonably could be assumed that, should the area be served by an independent fire district which mandated by its fire prevention code the provision of fire hydrants, the fire district could be liable for the hydrant fees.

Torts of Fire Fighters

The general rule is that a municipality is not liable for the torts of its fire fighters when they are performing a governmental function, when they act outside the scope of their authority, or where they are classified as public officers vested with discretionary immunity. Yet municipalities *are* liable for the torts of their fire fighters where statutes provide for it, such as during emergency vehicle operation, or where their acts create a nuisance, or when they are engaged in a nongovernmental function.

Other Fire Service Operations or Equipment

With respect to municipal liability, the following discussion deals with injuries arising from the negligent maintenance or utilization of fire fighting equipment other than fire apparatus, fire boats, or water distribution equipment.

Several cases have held that a municipality is not liable to pedestrians who stumble over fire hose, even though such negligence was predicated upon the failure to maintain the streets and sidewalks in a reasonably safe condition. As stated in *Powell v. Village of Fenton*, "If this necessary fire fighting apparatus crossing a sidewalk renders travel dangerous, temporary safety must yield to the more important and imminent demand for saving property and perhaps life by promptly suppressing the conflagration." This immunity probably would not exist, however, in a case where the hose stretched across the sidewalk was not being used in connection with fire extinguishing operations; as an example, where it was stretched out on the sidewalk in front of the fire station to be washed or dried.

In *Karczmarczyk v. Quinn*, a pumper of the Woonsocket, Rhode Island, Fire Department was being used to supply water to a construction project. Hoses were stretched across sidewalks to a supply hydrant and to the construction site. The plaintiff tripped over the hose which was lying in the gutter, not readily observable by pedestrian traffic. In holding the municipality liable, the supreme court stated: "The operation of a fire department is a governmental function, and a municipality is immune from liability for the negligent performance thereof. However, the acts of the firemen under consideration here were in no manner essential or relevant to the exercise of that function .... [Although] we recognized that the operation of a fire department as a governmental function requires more than the mere extinguishment of fires, ... the immunity from liability for the negligent performance of those functions arises only when the relevant conduct was in the discharge of duties essential to the performance of the governmental function."

When an unusual windstorm blew over an improperly constructed fire alarm tower, damaging the property of an adjoining owner (*Wilson v. City of Mason*), and when the wall of a fire cistern collapsed on an employee engaged in shoveling out silt (*Mulcairns v. Jonesville*), municipal liability was imposed.

Fire alarm signal wires also have been sources of liability to various cities. When a pole used solely for the support of a fire alarm box and signal wire broke off at the ground, falling on a boy playing in the street, a city was held liable. The court reasoned: "It is the duty of the city to exercise reasonable care to keep its streets free from danger, and it possesses the power to do whatever is necessary to accomplish this purpose. If, through negligence, it fails to perform this duty and an injury results from a dangerous condition of one of its streets, the city cannot avoid liability by showing that negligence in caring for an instrumentality used in the performance of a governmental duty is what brought about the dangerous condition." (*Hilstrom v. City of Saint Paul*)

Fire Stations and Premises Liability

Fire stations often are visited by children on school tours, by students of civics and social sciences, by fire fighters from other cities, by fire buffs, and by others out of mere curiosity. Therefore, it may be well to consider the city's responsibility to these persons for any injuries they might receive due to defects in the buildings or premises. The determination of municipal liability for fire department premises depends upon that state's determination of the nature of the activity—governmental or proprietary—and the nature of the duty owed to the plaintiff. The duty is determined either by the status of the individual, invitee, licensee, or trespasser, or the universal "reasonable person" standard of due care (see Chapter 8) as adopted by some jurisdictions. (*Rowland v. Christian*)[12]

A critical focus of the court will be whether or not the fire department owned, occupied, furnished, or controlled the premises. In *Denton v. Page*, a city initially was held liable for the injuries of an innocent third party occurring at the site of multiple arson fires involving a barn. The barn was neither owned nor occupied by the city, but was the subject of continuous surveillance by the city's fire investigation unit. The plaintiff was injured during the course of discovering another incendiary fire within the barn which was fueled by illegally stored flammable liquids. The trial and intermediate appellate courts held the city liable for the plaintiff's injuries based upon the city's failure either to warn the individual of the known hazards or to remove the hazardous materials. In reversing the appellate and trial court, the Texas Supreme Court stated that the city could not be liable as it did not own, occupy, or control the premises, nor had it created the dangerous condition. Surveillance of the site and constructive knowledge of the hazards therein did not give rise to a duty, on the part of the city, to correct the hazards in nonowned or controlled property or to warn all who might be affected.

---

[12] See *Dougherty v. Town of Belmont* for discussion of governmental immunity and governmental proprietary activities in operation of government facilities (such as parking lots).

See also *Coxson v. Gerry Volunteer Fire Department*. Here the fire department owed a very high duty to a paying guest on the defendant's premises, who was struck by a runaway rodeo horse.

Again, classification of the fire service function as governmental over ministerial can serve to limit liability. For example, with respect to pole hole injuries, in jurisdictions holding that the maintenance of a fire station is a governmental function, there is no municipal liability for injuries to members of the public who fall through such openings. In *Ford v. City of Caldwell* a child, visiting his father who was an on-duty fire fighter, fell through the pole hole. The state supreme court held the city not liable, as maintenance of the fire station was considered an integral element of the municipality's fire department, a governmental function which was a protected activity under governmental immunity. In *Nicastro v. Chicago* a child was taken into the station by a fire fighter who had no authority to do so. When the child fell through the pole hole, the fire fighter, but not the city, was held liable. It was held that the negligence of the fire fighter, and not a defect in the building, caused the injury. In *Barnes v. Waco* a doctor, who was called to examine a fire fighter applicant for an insurance policy, fell through a pole hole and was killed. The court held that the doctor was a mere licensee to whom the city owed no duty to keep the property in a safe condition.

On the other hand, it has been asserted by some jurisdictions that in the care and management of a fire station a city is performing a ministerial duty. Thus it is liable for injuries caused by defects of which it has had notice and reasonable opportunity to repair. That result was reached in *Walters v. City of Carthage*, where an apparatus door fell on a three-year-old boy playing on the sidewalk. The court held: "A municipal corporation is performing a ministerial duty in maintaining a fire station and is liable in damages for neglecting to make the same safe."

In some jurisdictions, liability may be imposed by statute on a municipality for injuries caused by defects in a fire station and its premises. Such states typically would impose liability for injuries to persons and property resulting from the dangerous or defective condition of government-owned or -operated buildings, grounds, and property. The municipality would be liable where the council or other entity having authority to remedy a dangerous or defective condition had knowledge or notice (actual or constructive) of it, yet failed or neglected to remedy the situation within a reasonable length of time, or to protect the public against it.

In *Boothly v. Town of Yreka City* the city had lent the county, free of charge, an upstairs room of the engine house for use as a polling place. The plaintiff's heel caught on a nail that was sticking up on a poorly lighted stairway, causing her to fall. In holding the city and county liable, the court said that the city owed a duty to the plaintiff, as a third party, to see that the premises were reasonably safe. The court added that insofar as the tenant fire department had knowledge of the defect, both the landlord and tenant were liable. Whether or not the tenancy was for compensation was immaterial.

Extra precautions should be taken in conducting station tours for children. For example, when a fire captain at the Central Fire Station in Baton Rouge,

Louisiana climbed into the cab of a fire truck to demonstrate the siren, he warned everyone to step back. However, he could not see a little nine-year-old girl stick her finger in the siren. When he pushed the button, the siren's blade amputated part of her finger. Since the child's mother, aunt, and seven cousins were present at the time, and since the captain's testimony that he had given them all warning to stand clear was believed by the trial court, the city was not held liable.[13]

Is a Fire Station an "Attractive Nuisance"?

Is a fire station a nuisance? Under the Attractive Nuisance Doctrine, a child who—because of his natural innocence—is tempted to play about dangerous machinery he finds fascinating is treated by the law as an invitee rather than a trespasser. This establishes a duty to use due care to prevent the child from being injured. Particular care should be exercised with respect to children who are frequent callers at fire stations. They sometimes slip by the fire fighter on watch and, unnoticed, wander around the apparatus floor or other parts of the station. It may be no defense to the city that the injured child was a trespasser if the court holds that it is a proper case for the application of the Attractive Nuisance Doctrine. Courts seem to make the liability rest upon the allurement of the nuisance, as well as upon the defendant's need to anticipate danger to children. The elements necessary to apply the doctrine are that a condition must be:

1. novel in character;
2. attractive and dangerous to children; and
3. easily guarded and made safe.

Certain fire apparatus and its equipment are no less attractive to children. Hence fire fighters should be particularly careful to safeguard children from possible injury.

Fire Department Facilities as Private Nuisances

A private nuisance is essentially an activity of a defendant that interferes with the peaceful use and enjoyment of a defendant's land. "The ownership or rightful possession of land necessarily involves the right not only to the unimpaired condition of the property itself, but also to some reasonable comfort and convenience in its occupation."[14] In essence, the utility of the defendant's conduct must be weighed against the amount of harm to the plaintiff, the question being not simply whether the plaintiff is annoyed or dis-

---

[13] Morse, "Legal Insight: Accident in a Fire Station—Negligence?" *Fire Command*, Vol. 40, No. 5, May 1973, pp. 48-49

[14] Prosser, *op. cit.*, p. 591

turbed, but whether that annoyance or disturbance arises from an unreasonable use of the neighbor's land or operation of its business. (*Sans v. Ramsey Golf and Country Club*)

For example, in *Gilmour v. Green Village Fire Department*, the court enjoined the fire department from using its playing fields for more than three night baseball games a week, and from using arc lights after certain hours, due to their effects on neighboring property holders. In another case, where the volunteer fire department's siren alerting system emitted sounds exceeding 128 decibels within 50 feet of adjacent residences, a court held that affected property owners could maintain a nuisance action against the fire department for operation of the siren alerting system. The harm to the plaintiff was judged to be the unreasonableness of the continued use of the siren system, when effective alternative means to alert the volunteers existed. (*Malhame v. Borough of Demarest*)

Destruction of Buildings to Prevent the Spread of Fire

At common law, a municipality needs no statutory authorization to destroy buildings to prevent the spread of fire, and is not liable for the destruction of buildings or their contents under such circumstances. This is based upon the theory that the greater public necessity to stop a developing conflagration can justify the destruction of private property. The theory has been described as follows: "Where the danger affects the entire community, or so many people that the public interest is involved, that interest serves as a complete justification to the defendant who acts to avert the peril to all. Thus, one who dynamites a house to stop the spread of a conflagration that threatens a town, or shoots a mad dog in the street, or burns clothing infected with small pox germs, or, in time of war, destroys property which should not be allowed to fall into the hands of the enemy, is not liable to the owner, so long as the emergency is great enough, and he has acted reasonably under the circumstances. The 'Champion of the Public' is not required to pay out of his own pocket for the general salvation. The number of persons who must be endangered in order to create a public necessity has not been determined by the courts."[15] The power to destroy buildings also carries with it the power to determine the necessity of so doing.

Various grounds for the right to exercise this power have been employed, including providing for the public welfare, public necessity, police power, abatement of a nuisance such as a house on fire, statutes, and eminent domain. In jurisdictions that follow the latter theory, the city will be liable to

---

[15] Prosser, *op. cit.*, Section 24, cited with approval in *Steele v. City of Houston*

compensate for the loss. This is because the taking under eminent domain carries with it the correlative duty to pay just compensation.

In *Steele v. City of Houston*, a house set afire during recapture of escaped convicts, who were hiding there, was held a taking, and the city was liable to recompense the homeowners for their losses. The burning of the house was considered a public convenience, to facilitate the police department's recapture of the escaped convicts. The property was taken for a public purpose, and not destroyed for public necessity in response to a major community-wide peril, such as a conflagration and because Texas statutes provided that a person whose house is destroyed for public convenience may recover under certain conditions.

Riots

Fundamental Considerations

Fire departments sometimes are called upon to assist law enforcement agencies in quelling riots and public disorders. Where there are no fires involved, the fire chief justifiably should have serious reservations about tieing up equipment on one side of town, for mob control, when at any moment a fire might break out on the other side of town. The chief also may be concerned if the fire department operates under statutory or charter provisions which limit its authority to preventing, controlling, and extinguishing dangerous fires.

Even under the assumption that fires are involved, what action can a fire department take when it is physically obstructed from access to the fires? Or placed under attack by rioters while attempting to control such fires?

Definition of a Riot

It is important, for several reasons, to know just what acts are considered sufficient to constitute a riot. What effect might a riot have on the authority of the fire chief to take unusual measures to cope with rioters? On direct performance of fire suppression activities? On the chief's possible liability for acting beyond his scope of authority?

The Federal government defines riot as "a public disturbance involving 1) an act or acts of violence by one or more persons part of an assemblage of three or more persons which act or acts shall constitute a clear and present danger of, or shall result in, damage or injury to the property of any other person or to the person of any other individual or 2) a threat or threats of the commission of an act or acts of violence by one or more persons having, individually or collectively, the ability of immediate execution of such threat or threats, where the performance of the threatened act or acts of violence would constitute a clear and present danger of, or would result in, damage or injury to the property of any other person or to the person of any other individual."[16]

Fire Department's Participation in Riots

From a legal standpoint, the role of the fire department in civil disturbances is fire control—not riot control. This is not merely a matter of tradition or personal preference on the part of fire department administrators. Rather, it is a matter of law. As pointed out earlier, the authority of any city fire department stems from the city charter. While the charter usually distinctly places upon the police department the duty to quell riots, the charter seldom even mentions riots in connection with the duties of the fire department whose primary job is concerned with fire prevention and control. It would be an *ultra vires* ("beyond the scope of the powers of a corporation") act for fire department members to engage in any mob control activity, except as might be incidental to the suppression of dangerous fires, or in self defense.

It is hard to conceive that any jury would find that fire fighters were not using reasonable diligence in fighting a fire merely because they were forced to withdraw from a fire area due to physical attack (including gunfire) where there were insufficient police or soldiers to provide a measure of protection. It certainly would seem that the only practical action when fire fighters are being attacked under these conditions would be to pull back to a perimeter where they could protect exposures. They would attempt to move inside the riot zone only when it was imperative to rescue people in danger of being trapped by fire there.

For example, in *Biloon's Electrical Service v. City of Wilmington*, the plaintiff repair shop owners brought suit against the city for negligence. Charges were inadequate fire protection and negligent fire suppression due to delayed fire department response. The fire occurred during a period of civil disturbance within an area where the police described citizen temperament as at the "breaking point." Upon hearing this assessment, the chief on duty advised police that fire equipment would not be dispatched for small trash fires in the area nor would equipment respond until fires were confirmed by police. In addition, police would have to escort all fire fighting equipment to the fire location.

A neighbor to the plaintiff's property, who was monitoring police department communications on her home scanner, observed the growing fire. Viewing its quick development with alarm, and noting the fire department's nonarrival, she repeatedly called the fire department to advise it of the serious nature of the emergency. Only after the on-duty chief arrived on the scene, approximately 58 minutes after the initial fire report, was fire apparatus dis-

---

[16] 18 United States Code Annotated Section 2102(a)

patched to the scene. Upon their arrival, "... fire fighters were accosted by a hostile crowd which pelted them with bricks and bottles and prevented access to hydrants. Not until additional units of police were called in could hoses be connected to the hydrants; however, the crowd continued its bombardment and the police used tear gas to disperse the rioters." The fire totally destroyed the plaintiff's structure.

The court held that the decision to delay response until the nature of the fire was confirmed, and to delay suppression activities until police reinforcements arrived, were protected discretionary decisions by the fire chief. "The rule is liability not immunity but exceptions to liability are necessary when public policy considerations far outweigh the value of an individual property. For instance, if under riot conditions snipers are shooting at police and firemen who respond to scenes of violence and arson, any discretionary decision to restrict municipal response to the conflagration for the purpose of saving lives must take precedence over the value of mere property. Any argument that attempts to question the merits of that decision based upon a full and complete record of the exigent circumstances presented to those who made the decisions leads to an attempt to second guess executive decisions. The danger in that process is the jeopardy of impairing the governing process itself." (*Biloon's v. Wilmington; Westminster Investing v. Murphy; Asian Consulting Laboratories v. City of Saratoga Springs*)

In such cases, it is important to show that a policy decision consciously balancing risks and advantages actually took place, and that the decision not to respond was not discriminatory but based upon the clear and present danger of civil disturbance in the area. (*King v. City of Seattle*)

A final word of caution: the carrying of a concealed weapon is a felony. Fire fighters should not carry a weapon, nor attempt to carry out police duties, unless empowered by law to do so.

Review Activities

1. Under what circumstances could a 911 emergency dispatch center be held liable for the personal injury or property damage of a citizen? What are the principal factors utilized by the court in considering the issue of liability?
2. To exercise the privileges of an emergency vehicle, what prerequisites/circumstances must have occurred? With an emergency vehicle, what additional privileges does the driver have, and what responsibilities?
3. Describe in your own words why the fire department was found liable for the death of Mrs. Schiable. Do you believe that the court reached a just conclusion? Why?
4. What is a "special relationship" as determined by a court and what is the significance of such a finding? What are the factors that a court

would consider in finding the existence of a "special relationship" between an individual and the fire department?

5. Give examples of how fire hoses, fire alarm signal wires, or fire hydrants could be considered sources of liability for a city.
6. What constitutes a riot? Describe the role played by the fire department during a riot.

Chapter Seven

Safety

Parties to Fire Fighter Safety

Local Government's Responsibility for Safety

A local government's responsibilities to its employees are similar to those of any other public or private employer. Such responsibilities are outlined in the laws governing master-servant relationships. As stated in *Taylor v. Hostetler*: "A duty rests upon the master not to expose the servant, in the discharge of his duty, to perils and dangers against which the master may guard by the exercise of reasonable care. These duties are: 1. To provide safe and suitable machinery and appliances for the business, including a safe place to work. This includes the exercise of reasonable care in furnishing such appliances, and the exercise of like care in keeping the same in repair and in making proper inspections and tests. 2. To exercise like care in providing and retaining sufficient and suitable servants for the business, and instructing those who, from newness or age, evidently need it. 3. To establish proper rules and regulations for the service and having adopted such, to conform to them."

The duty to provide a safe working environment is not limited to fire apparatus and fire stations. For example, in *Barger v. Mayor and City Council of Baltimore*, plaintiffs were fire boat crew members who suffered significant hearing loss due to diesel engine noise. Following their disability retirement, they sued the city for negligence by maintaining an unsafe working environment aboard its boats. In sustaining the trial verdict against the city, the court noted that it has long been recognized that prolonged exposure to loud noise can cause hearing loss. The city was deemed negligent by not undertaking remedial efforts or providing additional protective equipment.

The duty of the municipality to provide a safe workplace extends to the benefit of those assisting the fire department and to unauthorized individuals who respond with a fire department. In *Wolf v. City of New York*, an unauthorized fire buff responded with city apparatus to the scene of a building fire. With the knowledge of the company officer, the buff assisted fire fighters on the roof of the fire building. In response to the officer's directive, he left the roof via a route of limited visibility, over a parapet. He was seriously injured when he fell into an air shaft. The New York Court of Appeals, in sustaining the award to the injured buff, cited the trial court's jury instructions outlining the applicable law: "Did the defendant [the city] through its agents conduct itself as a reasonable prudent person under those circumstances? It matters not that there is a departmental rule or regulation against acts or omissions. . .. The issue here is whether those acts or omissions were in fact created. The mere fact that there are regulations against them [fire buffs riding apparatus and assisting fire fighters] does not absolve the defendant city of any negligence."

As outlined in Chapter 5, the Occupational Safety and Health Act (OSHA)[1] as originally promulgated was not held applicable to public fire departments.[2] However, where states elected to be the OSHA enforcement agents within their states, application of the OSHA standards, or local equivalents, was a condition of their assumption of responsibility. In adopting the Federal regulations, many states have expanded their scope to include fire fighters in formerly non-OSHA-protected positions.[3] Other states have encouraged voluntary compliance with OSHA requirements but have prohibited state agencies from enforcing OSHA provisions with civil or criminal penalties against local government entities.

Some states have a Public Liability Act which outlines a municipal government's liability for its facilities, equipment, employees, and agents, as well as the procedure for obtaining recovery for injury. Where such a statute is in effect, a municipality may be held liable for injuries resulting from defective

---

[1] The general duty clause of the Occupational Safety and Health Act requires each employer to: ". . .furnish to each of his employees employment and a place of employment which is free from recognized hazards that are causing or are likely to cause death or serious injury. . .[and] comply with occupational safety and health standards promulgated under this chapter." 29 United States Code (USC) Section 654(a) (1976)

[2] 29 USC 651 et seq.

[3] Guidelines in this area are contained in NFPA 1500, the Standard on Fire Department Occupational Safety and Health Program, adopted by the National Fire Protection Association in 1987. The document contains minimum requirements for organization and structure of an OSH program for fire suppression, rescue, and other emergency services. Provisions of NFPA 1500 are intended to help prevent accidents, injuries, and exposures; to help lessen the severity of these health problems when they do occur; and generally to reduce the probability of occupational fatalities, illnesses, and disabilities affecting personnel of public, private, and military fire departments and fire brigades.

fire apparatus. However, can a city be held liable for fire fighter injuries due to the negligent condition of new apparatus, such as unsafe apparatus built in accordance with a department's specifications? In general, as in *Smith v. F.W.D. Corp.*, where a city only provides the specifications for a piece of apparatus, and does not "... design the truck or occupy any role in the chain of distribution other than ultimate consumer, it is not liable in strict product liability for a design defect."

Again, in states having a Public Liability Act, the municipality may be held liable for injuries resulting from failure to provide effective supervision on the fireground. For example, provisions of New York's Public Liability Act state: "Section 2. Employer's liability for injuries. When personal injury is caused to an employee who is himself in the exercise of due care and diligence at the time .... 2. By reason of the negligence of the employer intrusted with any superintendence or by reason of the negligence of any person intrusted with authority to direct, control or command any employee in the performance of the duty of such employee, the employee ... shall have the same right of compensation and remedies against the employer as if the employee had not been an employee of nor in the service of the employer nor engaged in his work." Thus, where a fire fighter was injured when struck by a smoldering sofa thrown from an upper level of a fire building, recovery would have been possible based upon "... department's failure to have superior officer direct traffic in the jettison of items such as a couch or sofa." (*Lawrence v. City of New York*)

In the case of *MacClave v. New York*,[4] the appellate court upheld a jury verdict of $145,000 damages to the widow of a fire lieutenant. The officer died from carbon monoxide poisoning while wearing a filter-type mask in a smokey apartment fire. Of special importance is the fact that the city was found negligent on two counts: 1. because, a few months before the lieutenant died, the fire department had issued the filter-canister masks "for general fire fighting purposes"; and 2. because no means had been provided to enable fire fighters to determine the carbon monoxide level in the atmosphere where the masks were used. The court held that the city was negligent in furnishing these masks, making them available for use at fires, and, in effect, encouraging their use when their limitations were apparent. The court declared: "The city in furnishing masks which could not be used under any circumstances was guilty of negligence as a matter of law ...."

Another part of the court's comments has particular bearing on the use of breathing apparatus in atmospheres for which the equipment was not designed. The court stated: "We must agree with the general proposition advanced by the defendant that when an appliance, furnished by an employer, is used for a task for which it is not intended, the employer is ordinarily not

---

[4] "Legal Decisions on Filter-Canister Mask," *Firemen*, Vol. 33, No. 5, May 1966, p. 9

liable for injury resulting from such use, and so, if this mask were used for purposes other than fire fighting, for example skin diving, that principle would apply . . . ."

Union's Responsibility for Safety

Safety as an Issue of Collective Bargaining

Under the various public employee relations laws, fire fighters are empowered to organize collective bargaining units for management-employee determination of terms and conditions of employment. As construed by the National Labor Relations Board and the various state boards, employee health and safety is considered a mandatory issue for collective bargaining.[5]

A public employer is not required to negotiate the substance of its public mission, the methods it selects to fulfill that mission, or similar issues which may affect terms and conditions of employment. However, it can be compelled to negotiate about the impact of such decisions on employment terms and conditions.[6] Thus, as outlined in Chapter 5, minimum apparatus manning generally is not considered a mandatory issue of collective bargaining, but the impact of the manning provided on an apparatus, a piece of equipment, or at a fire scene can be. Such impact can result in demands for hazardous duty pay in response to diminished unit manning and the attendant increased risk of fire fighter injury.[7] Other issues affecting public employee health and safety that have been considered mandatory subjects of collective bargaining include employee physical fitness programs, medical review boards, and adoption of National Fire Protection Association (NFPA) standards on fire fighter safety.

Union's Liability for Failure to Perform Safety-Related Duties

Public employees have sought recovery for injuries from unions based upon two theories: 1. breach of the duty of fair representation; and 2. state law negligence in the performance of safety duties assumed under a collective bargaining agreement. In general, suits based upon breach of a duty of fair representation have been unsuccessful. However, those based upon failure to perform safety duties assumed under a collective bargaining agreement have had some success.[8]

---

[5] 29 USC Sections 151-169 (1976)

[6] Leonard, Arthur S., "Collective Bargaining on Issues of Health and Safety in the Public Sector: The Experience under New York's Taylor Law," *Buffalo Law Review*, Vol. 31, 1982

[7] Fire Fighters Union Local 189, 11 New York P.E.R.B. 3087 (1978)

[8] See Segall, Tony, "The Wrong Pocket: Union Liability for Health and Safety Hazards," *Industrial Relations Law Journal*, Vol. 4:390, 1981.

Fire Equipment Manufacturer's Responsibility for Safety

Under theories of breach of expressed or implied warranty, negligence, or strict liability for defective products, fire equipment manufacturers have been held liable for fire fighter injuries. Refer to Chapter 8 for an outline of these legal theories. In sustaining a fire fighter's recovery for injuries, the court's critical focus will be on whether or not the equipment was warranted to be in good condition and was used properly and as intended; and whether or not the fire fighter's injuries resulted directly from failure of the equipment.[9]

Limitations on Liability

Assumption of the Risk

Assumption of the risk generally bars recovery for damages by an injured employee who knew of the danger in a situation but nevertheless voluntarily exposed himself to it. The policy basis for this limitation on employee recovery is that an employer should not be held liable for injuries knowingly and voluntarily risked as an inherent part of the job. Risks that are normally incident to a job, that are not unusual, and of which the employer possesses no greater knowledge should not serve as bases of employer liability.

"'Assumption of risk,' in the law of master and servant, is a phrase commonly used to describe a term or condition in the contract of employment, either express or implied from the circumstances of the employment, by which the employee or servant agrees that certain dangers of injury, while he is engaged in the service for which he is hired, shall be at the risk of the employee or servant." (*Blackmore v. Auer*)

To preclude recovery by the employee, assumptions of risk cannot be in response to an employer's negligence. An employer's actions must be free of neglect in order for the rule to apply. In *Jackson v. City of Kansas City*, two fire trucks from the city's fire department collided en route to a reported building fire, resulting in significant personal injuries. The city contended that fire fighters assume the risk of collisions when they respond in a fire truck. However, the court held that such collisions are extremely rare and thus, as a matter of law, it is not reasonable to believe that fire fighters knowingly assume such risks.

Fellow Servant Rule

A second limitation on the right of a fire fighter to recover against a city is the common law "Fellow Servant Rule." A highly discredited legal theory, it still remains valid in some jurisdictions. The New York Court of Appeals has summarized the rule as follows: "If the co-servant, whose act caused the

---

[9] See "Products Liability: Firefighting Equipment," 19 *American Law Reports* 4th 310.

injury, was at the time representing the master in doing the master's duty, the master is liable. If, on the other hand, the co-servant was simply performing the work of a servant, merely in his character as a servant or employee, the master is not liable. Moreover, the fact that the person whose negligence caused the injury was a servant of a higher grade than the servant injured, or that the latter was subject to the direction of the former, and was engaged at the time in executing the orders of the former, does not take the case out of the operation of the general rule, nor make the master liable." (*Loughlin v. State of New York*)

In the *Lawrence* case discussed earlier, involving the jettisoned sofa which injured the fire fighter, the court held that the Fellow Servant Rule was applicable, limiting city liability. Although the city had a nondeligable duty to provide a safe working environment, it had no obligation to control all activities of all the fire fighters. As admitted by the injured man, it was not unusual to simply warn fire fighters of forthcoming dumping of debris during overhaul, rather than having an officer direct the jettisoning of falling objects. As aptly stated: "In many cases there are no fixed conditions; the danger waxes and wanes during the progress of the work, and the safety of the group of workmen depends upon the care with which other workmen in the same group perform their work. If a master has properly performed the duties of management, he is not responsible for the temporary conditions of unsafety which arise during the progress of the work."[10]

Defective Streets

A municipal corporation is liable for injuries to fire fighters caused by defects in streets, even if the injuries occur during performance of a governmental function. In one case, a fire fighter was jolted off the apparatus when it struck a hole in the street. The injury was found to be caused not by the negligence of the driver in striking the hole, but by the city's negligence in not repairing the hole. In another case, the apparatus driver was killed when his truck hit an unguarded and unlighted pile of building materials left in the street with the permission of the city. The court declared that cities are required to keep and maintain their streets in a condition reasonably safe for public travel. The cities are held to as great a degree of care toward a fire fighter/driver using the streets in the discharge of his duties as they are to any other traveler.

Review Activities

1. Explain why, in those states having Public Liability Acts, a municipality may be liable for personal injuries resulting from inadequate supervision of the fireground.

---

[10] Restatement, Agency 2d, Section 500. Cited with approval in *Lawrence*, supra.

2. Can a city be responsible for a fire fighter's injuries due to the negligent design of a fire truck? Explain.
3. Explain the policy basis underlying the Fellow Servant Rule. Do you consider it still valid? Explain.
4. State whether or not your department is subject to the Occupational Safety and Health Act standards. Provide examples of areas where OSHA standards are directly applicable to the performance of fire department suppression services.

Chapter Eight

Fire Prevention Bureaus

The Legal Basis for Code Enforcement

In transmitting the report *America Burning* to the President, Chairman Richard E. Bland of the National Commission on Fire Prevention and Control stated the Commission's recommendations for improving the nation's fire safety. He wrote, in part, ". . . emphasize prevention of fire through implementation of local programs. This is in keeping with the very nature of the fire problem which is felt hardest at the community level." This recommended local emphasis on controlling "fire in the built environment" consisted of improved efforts in three areas: fire protection engineering; public education; and fire safety regulation and enforcement.[1] The first two points are voluntary efforts and as such do not require utilization of the powers of the state, through exercise of the law, to be successful. The third point, fire safety regulation and enforcement, does require such exercise. The source, basis, nature, and extent of such power, as well as its limitations and potential liabilities, is the focus of this chapter.

Police Powers in General

Police power is the fundamental power of the state to place restraints upon the personal freedom and property rights of persons for the protection of the public health, safety, and welfare, and for the promotion of the public convenience and general prosperity. (*Marshall v. Kansas City*) The police power

[1] *America Burning*, Report of the National Commission on Fire Prevention and Control, 1973

is as broad and extensive as the need for safeguarding the public interest. (*Land v. Mt. Vernon*) The police power is an inherent power of the individual states, possessed by them before adoption of the United States Constitution and reserved to them under the Tenth Amendment.

Municipalities, however, have no inherent police power and can exercise only such police powers as are vested in them by the states under constitutional provisions, statutes, or charters.[2] Certain cities have been delegated police powers fully as broad as those which the states themselves exercise. Generally, municipalities are endowed with broad and rather general police powers to cope with the problems of health, welfare, morals, and public safety within their respective corporate boundaries.[3]

Distinguished from a specifically enumerated power of the state, such as eminent domain, police powers are residual powers. These powers encompass all that are reasonably necessary to carry out the purpose of government. "This police power of the state extends to the protection of the lives, limbs, health, comfort, and quiet of all persons and the protection of all property within the state." (*Thorpe v. Rutland Railroad*) This power has been amplified even more by the United States Supreme Court which stated: "We hold that the police power of a state embraces regulations designed to promote the public convenience and general prosperity, as well as regulations designed to promote the public health, the public morals, or the public safety." The extent of the police power is exemplified by the government's authority to destroy private property or regulate its use to such an extent that the property's value is materially impaired, provided the property is not actually taken for use by the public agency for its own purposes.

The police power to regulate extends to all individuals, corporations, associations, and other entities engaged in business, professional, or any other pursuit or activity within the community.[4] The scope of that power is limited only by the bounds of the state grant to the community and the constitutional protection provided each individual by both the state and Federal constitutions. "[A] municipality can exercise only so much of the police power as may be expressly granted or necessarily implied from those powers expressly conferred by the [legislature] or those necessarily implied from those ex-

---

[2] The often-quoted Dillons's Rule defines the generally recognized limits of local government: Local governments have 1. those powers expressly conferred by state constitution, state statutes, and (where applicable) home-rule charter; 2. those powers necessarily and fairly implied in, or incident to, the powers expressly granted; and 3. those powers essential to the declared objects and purposes of the municipality or quasi corporation. Reynolds, Osborne M., *Hornbook of Local Government Law*, West Pub. Co., 1982, p. 137

[3] Rhyne, Charles S., *The Law of Local Government Operations*, National Institute of Municipal Law Officers, 1980, p. 447

[4] Rhyne, *op. cit.*, p. 451

pressly conferred; if the action taken exceeds that power, the action is void." (*Baltimore Planning Commission v. Victor*) It is within this context of delegated police powers that local government and the fire department must operate.

Police Powers and Fire Prevention Regulation

From time immemorial, man has recognized the fact that uncontrolled fire is a menace to society. Through exercise of police powers, certain restrictions were imposed on the conditions of fire's creation and use by mankind. At first, laws in the field of fire safety were general in nature, for it was considered common knowledge that careless conduct with respect to fire could lead to disastrous results. Soon, however, it was obviously not enough to merely lay down the rule that "no person shall create a fire hazard"; no one could reasonably be expected to know, under all circumstances, every condition that creates a fire and what preventive measures are appropriate. As a result, rules were promulgated to advise the public what specific actions and conduct were intended regarding fire.

But what is the extent of local government's authority, under the police powers, to regulate in the interests of fire safety?

Laws governing the conditions that give rise to dangerous fires are subject only to the ordinary limitations on the exercise of the police power. Justice Shaw in 1853 gave an early explanation of the application of the police power to individual property owners: "We think it is a settled principle, growing out of the nature of well-ordered civil society, that every holder of property, however absolute and unqualified may be his title, holds it under the implied liability that his use of it may be so regulated, that it shall not be injurious to the equal enjoyment of others having an equal right to the enjoyment of their property, nor injurious to the rights of the community." (*Commonwealth v. Alger*)

The police power frequently has been used to promote fire safety in various fields of legislation. Better fire protection has been one of the major grounds in upholding the constitutionality of zoning and set-back ordinances; limitations on the height, bulk, and area of buildings; restrictions for the location of roadways and garages; etc.

Thus, a municipality has ample authority through the exercise of its police power, as well as through the rights granted by its charter, to deal with arson, incendiarism, nuisances, building construction, location of hazardous industries, occupancy, and numerous other important factors relative to the prevention of fire. The only general limitations are that such regulations and activities must not be arbitrarily exercised or applied discriminatorily, and must bear a reasonable relationship to the fire safety objectives being sought.

Fire safety regulation requires that an entity possess the authority to promulgate—as well as the power to enforce—specific requirements. It is essential to know what authority lies with the fire safety agency or fire department. Reference must be made to authority given to the state regulatory agency by statute; to the fire department by charter, statute, or ordinance; and to the specific powers given to the bureau of fire prevention.

For example, the State Fire Marshal of California is directed by statute to prepare and adopt regulations establishing minimum requirements for the prevention of fire and for the protection of life and property against fire and panic in any building or structure used or intended for use as an asylum, jail, mental hospital, hospital, sanitarium, home for the aged, children's nursery, children's home or institution, school, theater, dance hall, skating rink, auditorium, assembly hall, meeting hall, night club, fair building, or similar place of assemblage, etc.[5] Thus, the state fire authority is granted the specific, though potentially not exclusive, authority to regulate for fire safety in these occupancies. Without such enumerated authority, all the myriad regulations adopted by that state's fire marshal would be *ultra vires* or beyond the scope of his authority and unenforceable under the law.

Under provisions of the City of Los Angeles' charter, the fire department "shall have the power and duty to control and extinguish injurious or dangerous fires and to remove that which is liable to cause such fires, and to enforce all ordinances and laws relating to the prevention or spreading of fires, and all ordinances and laws pertaining to fire control and fire hazards within the City of Los Angeles . . . ." In addition, municipal ordinances of Los Angeles authorize the chief of the fire department to make and enforce such rules and regulations to prevent and control fires and fire or explosion hazards as are necessary to carry out the purposes of the fire safety ordinances.

The power granted by the Los Angeles charter extends to enforcement of *all* laws and ordinances relating to the prevention of fire. Hence, whether regulations are found in sections of the municipal code dealing with building requirements, or sections relating to health requirements, or even in the laws of the state (such as State Fire Marshal requirements), the fire department has the charter-given power to enforce these regulations if they relate to the preventing or spreading of fires. Thus, in Los Angeles and most cities, the fire department or its fire prevention bureau has the power to utilize every regulation and enforcement technique permitted by law to make the city safe from fire.

Even without such specific authorization in the charter or municipal ordinances, such power would exist as a reasonable exercise of the government's police power. The late Robert S. Moulton, when he was Technical Secretary of the National Fire Protection Association, made this statement which remains true today:

"Following the Cocoanut Grove fire, the officials concerned were quoted in the newspapers as stating that the existing codes and laws were not adequate

---

[5] California Health and Safety Code, Section 13143

to assure safety in places of public assembly such as this night club. We cannot believe, however, that this disaster [in which almost 500 lives were lost] is chargeable to any deficiency in the law.... In our opinion, building and fire officials can now do practically everything that is necessary to assure public safety from fire without any more laws.... Even though there may be no further authority than a city charter, these officials can issue orders for the purpose of safeguarding the public and in all probability have them upheld by the courts."[6]

In discussing the broad authority often delegated to a fire marshal via state laws and local ordinances, Judge Virgil Langtry of the Multnomah County Circuit Court in Portland, Oregon, stated: "I believe that a general delegation of authority to a fire administrator can be upheld, even though a finger cannot be pointed to a grant of power to do each specific thing, so long as the official does not discriminate between people or corporations in the same class, does not set up discriminatory classes which are not based on reason, that his action is reasonable as distinguished from being arbitrary, and his actions are directed toward achieving the desirable intentions of a fire code. If the fire marshal can confine his action within these boundaries, and he has an attorney who will represent the trend of liberalizing decisions with the logic which undergirds them, he may expect to take valuable action in the field of fire prevention, even though he does not find it specifically mentioned in his fire or building code."[7]

Several decisions have bearing upon the scope of the police power and its limitations as applied to fire prevention. Ordinances fixing the boundaries of fire districts have met little objection, and early cases have upheld the power to destroy buildings to prevent the spread of fire. A New York court asserted that "it has been repeatedly held that the fire commissioner has the authority under the ordinances to... order... the owner of a building used for manufacturing purposes to install an automatic sprinkler system to extinguish fires." (*People v. Miller*)

In another case, *Maguire v. Reardon*, the plaintiff was unsuccessful in his attempt to forbid demolition of a frame building after the city had given him a permit to install a large, expensive oven there. The United States Supreme Court found that the building had been unlawfully erected within the fire district, and that its demolition was not an unreasonable exercise of the police power; further, that the city could not be estopped for any inconsistent conduct because the police power cannot be contracted or bartered away; nor was there impairment of any obligation of contract that the owner might

[6] Moulton, Robert S., "Fire Inspection," *NFPA Quarterly*, Vol. 36, No. 3, Jan. 1943, pp. 197-198

[7] Langtry, Virgil, "The Fire Marshal's Delegated Authority," *NFPA Quarterly*, Vol. 57, No. 2, Oct. 1963, pp. 126-129

have had with the tenants, for all contracts are entered into with a reasonable exercise of the police power as an implied condition.

In a Montana case, *State ex re Brook v. Cook*, the facts were, briefly: the defendant owned a 40-year-old, two-story, frame dwelling in a residential district that was in general disrepair. The state fire marshal was empowered to inspect and, in proper cases, to condemn any building which "for want of proper repair, by reason of age, dilapidated condition,... or for any other cause or reason is especially liable to fire." Where it "is so situated as to endanger other buildings and property," the state fire marshal could declare it a public nuisance and order the condition remedied; but if the order is not obeyed, the fire marshal may not summarily proceed. He can only maintain an action against the owner "for the purpose of procuring an order from the court," along the same lines as the order given to the owner by him. The building in question, in addition to having been partly damaged by fire, was an "eyesore" to the neighborhood, although it was not located closer than 40 feet to the nearest adjacent building. A 20-day notice was given and ignored, so a complaint was filed setting forth the facts about the building and asking that it be declared a public nuisance, and either repaired to remedy its dangerous condition or torn down and removed. The Montana Supreme Court held the measures constitutional, but warned that "such a statute should be administered with caution and for the purpose of fire protection alone, and not to promote the 'city beautiful' idea." The court said, "We cannot overthrow a judgment based upon a preponderance of substantial evidence following the wording of the statute, even though we may suspect that the real purpose of the proceeding is to relieve the neighbors of a local 'eyesore,' or be inclined to believe with the witness for the defense, that the building is no greater fire hazard than it would be if repaired."

Elimination of Nuisances Constituting "Fire Hazards"

"Fire hazard" is but another name for "public nuisance," and the methods of removing both are the same. It is well settled that municipalities may enact ordinances declaring that certain conditions create a nuisance, and under its police power provide for abatement of the nuisance. (*Jones v. Odessa*)[8]

---

[8] Charter provision giving the city council power "to establish all necessary rules and regulations protecting the health" of the residents, and also "to define all nuisances and prohibit the same within the city" constituted broad grant of authority and was clearly adequate to permit enactment of an ordinance for destruction of buildings that were "fire hazards" or "dangerous to human life."

See also: Uniform Code for the Abatement of Dangerous Buildings, International Conference of Building Officials.

"Accordingly, the substance of local power to abate the nuisance is regarded as essentially a judicial question turning on whether or not a nuisance in fact is established." (*Hislop v. Rodgers*)[9] It has been held, for example, "that the state may order the destruction of a house falling to decay or otherwise endangering the lives of passers-by; the demolition of such as are in the path of a conflagration; the prohibition of wooden buildings in cities; the restriction of objectionable trades to certain localities." (*Lawton v. Steele*)

In *Kaufman v. Stein*, the court stated that a "wooden structure is not a 'nuisance per se.' It is the circumstances that make it a nuisance. Even when they are originally built in a place remote from the habitations of men, or from public places, if they become nuisances by reason of roads afterwards being laid out in that vicinity, or by dwellings subsequently erected within the sphere of their effects, the fact of their existence prior to the laying out of the roads or the erection of the buildings is no defense."

Care must be taken to allege sufficient facts to indicate to the court that a particular building constitutes a nuisance. Further from *Kaufman v. Stein*, "if such a building is a nuisance at all, it is a nuisance *per accidens*—because of its use, location, surroundings or other circumstances." Where several witnesses testified that "by reason of its dilapidated condition, loosened siding, curled shingles, decaying beams, being without a substantial foundation, with openings in the roof and debris inside, this building was especially liable to fire from outside causes, spontaneous combustion from damp and decay, vandals entering and lighting matches, mice and rats coming in contact with matches left in the building, and the like, and that if it did burn, by reason of its condition, sparks, burning shingles, and materials were likely to be carried much farther than the nearest buildings . . . ." Where expert evidence also was introduced to show that the building was in such a state of ruin and decay that it could not be repaired, the court found these allegations sufficient to describe a public nuisance that could be abated by an action instituted by the fire marshal.

Without expressly declaring a building to be a public nuisance, the trial court in California ordered it abated in an action instituted by the San Francisco City Attorney against the landowner. The appellate court said, "The trial court did . . . find . . . that said structure was old, dilapidated, unsafe, defective, abandoned, and extremely dangerous; that the chimney thereon was dangerous, extremely defective and in an unsafe condition; that practically the entire structure was rotten, broken, did not conform to building laws, and was a menace to life and limb of passers-by and to the property adjacent thereto; furthermore that it was an extreme fire hazard and unsanitary." Clearly the above findings warranted the trial court to direct that the structure be demolished.

---

[9] Sands, C. Dallas and Michael Libonati, *Local Government Law*, Callaghan and Co., 1986, Section 16.28

The authority to have a building declared a nuisance includes the authority to abate the nuisance. In *City of Boston v. Ditson*, the fire department, after more than seven years of complaints regarding accumulation of rubbish on the Ditson premises, issued an abatement order directing the premises be cleaned up within 24 hours. When the conditions remained, the fire department arranged for an independent contractor to clean up the site, and billed the owner for the costs. When the charges remained unpaid after the statutorily prescribed period, a tax lien was placed upon the owner's property, resulting in its forced sale. The court of appeals sustained the authority of the city, enforced through order of the court, to have the Ditson premises declared a nuisance and order the hazardous conditions abated. Although a 24-hour notice was considered very short, the court sustained the action. "She had had ample warning before the order was served that the condition could not be permitted to continue, and in any event a 'person owning a building in such shape had reasonable ground to expect that public authorities soon might require its prompt rehabilitation.' . . . "

Similar to the administrative warrant requirements for building inspections, execution of abatement orders must conform fully to the due process requirements of the United States Constitution. The abatement order, by itself, does not grant anyone the authority to enter premises to assure the order's enforcement. Rather, nonpermissive entry can be done only under the authority and by the direction of a valid search warrant. In the Ditson case, the fire department entered the premises without a search warrant, believing that the abatement order constituted exigent circumstances permitting warrantless entry. The court stated: "the record in the present case will not support either a finding of consent or of the type of emergency situation suggested by the [United States Supreme] Court to permit warrantless entries. The fire department had received complaints about accumulations of rubbish on the respondent's property for seven years before undertaking the cleanup in 1969. The abatement order itself, following the requirements . . . allowed 24 hours before the fire department could enter to clean up the premises. That waiting period belies any claim of exigency."

Exigent circumstances could exist where the condition of a building presents an immediate, direct danger to the public. In *D.F.P. Enterprises v. City of Waterbury* the plaintiff's building was destroyed by a fire in which 14 people died. Because of the dangerous condition of the burned structure and the need to retrieve the bodies, the city immediately arranged for private demolition efforts. The appellate court, in sustaining the mayors's authority to order immediate demolition, noted the broad emergency powers of Waterbury's chief executive; these powers are typical of those held by most municipal chief governing officials when confronted with a true emergency. By city charter in Waterbury, "the mayor shall have power: (a) To assume the entire control and direction of the police and fire forces of the city, or either of them, for a period not exceeding fifteen (15) days, at his discretion, in case of emergency, and to exercise all of the powers conferred upon the police and fire departments thereto.... In the event of a disaster emergency, the mayor shall assume full command of all local government functions and facilities and personnel. All available local resources shall be brought to bear on the emergency existing in an effort to protect and preserve human life and property of the community to the greatest extent possible. The mayor shall be limited only by the Governor of the State of Connecticut in state-declared emergencies or the President of the United States of America in a declared national emergency." By virtue of both of these sections, the mayor was authorized to assume full command of the situation that developed at the scene of the fire and to take all action necessary to protect and preserve human life and property. Given this broad grant of power, there is nothing in the record that would justify an intrusion by this court, with the benefit of hindsight, into the factual findings [existence of a dangerous public nuisance requiring immediate abatement] made by the trial court with regard to the exercise of these emergency powers."

The cost of the emergency abatement services were charged against the property owner and collected through enforcement of a lien on the property. The court sustained this cost recovery procedure, noting: "The lienable expenses are, in logic, those which are necessary to render safe fire-damaged real estate which poses a threat to public safety. That these expenses should be recoverable from the owner flows naturally from the duty which every property owner bears to maintain his property in such a way as will not occasion unnecessary damage or annoyance to others. If a municipality were not reimbursed for these expenses, the owner would be unjustly enriched."

In addition to actions that may be brought by public authorities, it is well to point out another possibility to the property owner whose building constitutes a nuisance: the civil liability of individuals who maintain a nuisance. In a Pennsylvania case, it was said: "It is true that a private person not specially aggrieved cannot abate a public nuisance, and especially where a statute provides a remedy for an offense created by it, that must be followed. It is well settled, however, that a private person, if specially aggrieved by a public nuisance may abate it.... If the owner or tenant of a powder magazine should madly or wickedly insist upon smoking a cigar on the premises, can anyone doubt that a policeman or even a neighbor could justify in trespass, forcibly ejecting him and his cigar from his own premises?" Statutes in many states grant anyone the right to abate a nuisance, where it can be done without causing damage or a breach of the peace.

Professor Westel Willoughby,[10] in one of his works on constitutional law, has the following to add on the subject of nuisances: "The aid of equity may

---

[10] Willoughby, Westel W., *Constitutional Law of the United States*, Baker, Voorhis Pub. Co., 2nd Student Edition, 1933

be invoked for the abatement of nuisances whether public or private, or in private nuisances, the individual may, at his own risk, himself undertake the abatement. In either case, whether by a process of the court or by personal action, notice to the person responsible for the nuisance is not required where there is an emergency. This constitutes an exception to the general principle that private property may not be taken or destroyed except after judicial proceedings and condemnation with notice to the owner."

Weed Abatement Programs

Any fire department which attempts to execute a weed abatement program is likely to have legal problems, not only in determining the owners to whom notices should be sent, but in other matters as well. If the owner fails to have the weed control work done within the prescribed period, the fire department—where authorized by law—may hire a contractor to do it. The contractor either bills the owner directly, or is paid from fire department funds; in the latter case the costs are added to the tax assessment for the land involved. Someone from the fire department should supervise the program to see that the right parties receive the notices, that only parcels which create a definite fire hazard are posted, that the same standards are applied throughout the community, that the right property owners are billed, and that valuable trees are not needlessly cut down. Never give an owner cause to demand payment for trees destroyed. In addition, if controlled burning is attempted by the owner, contractor, or fire department, all the conditions of the burn permit must be met; standby fire suppression equipment must be on the scene, manned and ready for action. Every precaution must be taken to prevent the fire from negligently escaping the bounds of the burn to damage adjacent premises.

Retrospective Measures

How far may a fire prevention bureau go in asking for installation of safety features, required by laws enacted since erection of the building? The answer focuses upon the scope and applicability of retrospective fire safety laws. "Retrospective building laws prohibit the future use of a building in certain ways unless changes are made. Retrospective or retroactive codes are not *ex post facto* laws. *Ex post facto* laws make past conduct criminal, while retrospective laws affect only future conduct."[11]

Some legislation expressly has been made retroactive, and its validity upheld. For example, it is commonly required that stairways and vertical shafts in buildings more than two stories high be enclosed, and that new means of egress and fire doors be installed on existing buildings even though they were

[11] Brannigan, Francis L., "Applying New Laws to Existing Buildings: Retrospective Fire Safety Codes," *Journal of Urban Law*, Vol. 60:447, 1983, p. 449

in accordance with the laws in existence when they were built. Some model codes include recommended provisions for retrospective application (commonly called "retrofitting") in high-rise buildings or special hazard structures.[12]

It has been held, however, that an ordinance cannot be made retroactive in regard to the removal of existing buildings from within fire area limits. Nevertheless, such structures may be barred from being moved into the fire limits.[13]

Ralph P. Dumont, an attorney in New London, Connecticut, summarized the problem of retroactivity as follows:

> A law is not unconstitutional, as a general rule, merely because it is retroactive in application or operation. There are only four basic situations in which retroactivity might be a proper reason for condemning a law:
>
> 1. An *ex post facto* law;
> 2. A law impairing the obligation of contract, but this refers only to state laws;
> 3. A law violating the Fifth or Fourteenth Amendment, because it takes property without due process of law; and
> 4. A law of a state violating the Fourteenth Amendment's requirement of equal protection of the law.
>
> *Ex post facto* refers only to criminal law . . . . Insofar as the second point is concerned, is hard to see how an existing building is a contract. We are left, therefore, with the problem of due process and of equal protection of the laws.[14]

Thus, legal analysis of the retrospective concept focuses on two constitutionally protected rights: property cannot be taken without due process of law; and similarly situated individuals cannot be denied equal protection of the laws. Where notice and an opportunity to be heard have been provided, it has been held that reasonable police power measures which increase the public safety and reduce the danger from fire do not violate these constitutional safeguards. This is true even though the measures may require changes in an existing structure, or reduce its utility.

Retroactive application of police power measures requiring the repair and improvement of buildings generally have been held not to violate the requirement of due process. The rule has been justified in these words: "There is no

---

[12] See: Uniform Building Code, 1985 Edition, International Conference of Building Officials, Appendix, Chapter 1, Division 1, Life Safety Requirements for Existing Buildings Other Than High-Rise Buildings, and Division 2, Life Safety Requirements for Existing High-Rise Buildings.

[13] Fire limits are geographical areas within an incorporated city wherein nonrated or combustible wood frame construction typically is prohibited.

[14] Dumont, Ralph P., "The Doctrine of Retroactivity and Fire Laws," *Firemen*, Vol. 30, No. 10, Oct. 1963, pp. 32-33

such thing as an inherent or vested right to imperil the health or impair the safety of the community.... It would be a sad commentary on the law, if municipalities were powerless to compel the adoption of the best methods for protecting life in such cases, simply because the confessedly faulty method in use was the method provided by law at the time of its construction."[15]

"Because fires and epidemics may ravage large urban areas, because unsightly conditions adversely affect the economic values of neighboring structures, numerous courts have upheld the police power of municipalities to impose such minimum standards even upon existing structures." (*Camara v. Municipal Court*)

Following the Triangle Shirtwaist Company fire of 1911, which resulted in the deaths of 145 workers within a "fire-proof" building, retrospective laws were passed to improve fire safety conditions within New York City's garment factories. The laws were challenged on due process, equal protection, and economic hardship grounds. In sustaining the retrospective laws, Judge Edward J. Gavegan of the New York Supreme Court stated: "It is true that these provisions constitute a forward step in remedial legislation quite in advance of anything heretofore undertaken by the legislature in the exercise of the police powers. It is also true that in many instances, clearly so in the case at bar, compliance with these provisions may entail hardship upon the owners of buildings which at the time of their erection fully complied with all existing provisions of law. Hardship, however, does not mean confiscation, and unless it results from the unreasonable exercise of arbitrary power on the part of the legislature no degree of hardship can justify the court in nullifying legislative enactments embodying the will of the people, since it is the primary duty of the legislature to protect the common interests of the whole people, even at the expense of personal or local interests. The mere fact that the law cannot be enforced without causing expense to the citizen who comes within its provisions furnishes no constitutional obstacle to such enforcement.... That the police powers of the state include everything essential to the public safety, justifying the destruction or abatement of public nuisances, even to the demolition of buildings in the path of conflagration, has long been settled law. The fact that the exercise of the power may disturb the enjoyment of individual rights without compensation for such disturbances does not make laws and regulations of a public nature unconstitutional, for they do not appropriate property for public use, but simply regulate its use and enjoyment by the owner.... If the extent of the injury which would be inflicted upon private individuals were controlling upon the constitutionality of the act, then it would be difficult, if not impossible, to have any fixed criterion, since the cost of compliance in some cases might be very

[15] Rhyne, *op. cit.*, Section 19.16

small, while in others it might equal, or even exceed, the equity of the owner. Such considerations are proper to address to the legislature, but not the court. (*Cockroft v. Mitchell*)"[16]

Following two major high-rise fires, the City of New York retrospectively required existing high-rise office buildings to provide "a fire safety plan to organize and effect evacuation from perilous areas without panic; an alarm/communications system, including elevator recall, to alert occupants to danger and to direct safe evacuation and effective internal firefighting; and, in centrally air conditioned buildings, either complete sprinklering, or both subdivision of open-floor areas with fire resistant partitions [known as compartmentation] and a system of smoke control to provide for safe evacuation." (*McCallin v. Walsh*) More than 870 structures would be affected by the regulations, necessitating costly capital improvements and adoption of time-consuming fire safety operating procedures.

The law was quickly and vigorously challenged. The trial court declared major portions of the law unconstitutional, holding that some options were technologically unsound, thus unfeasible, and many of the requirements were disproportionately costly for the benefits gained.

The appellate court reversed the trial opinion and sustained the retrospective high-rise fire safety improvements. The court stated: "Protection of the safety of persons is one of the traditional uses of the police power of the states. Experts may differ as to the most appropriate way of dealing with fire hazards. But the legislature may choose not to take the chance that human life will be lost and adopt the most conservative course which science and engineering offer. It is for the legislature to decide what regulations are needed to reduce fire hazards to the minimum. Many types of social legislation may diminish the value of the property which is regulated. The extreme

---

[16] "The state can enact remedial laws only after it has been demonstrated that there are evils to be remedied. Actual experience, even human sacrifice, is often required in order to arrest the attention of the public and shock it into a realization of the necessity for remedial laws. Experience in the matter of fire danger shows that so-called fire-proof structures sometimes burn;... that automatic fire alarms sometimes fail to alarm; that inside fire apparatus sometimes proves ineffective; that smoke from flammable contents of fire-proof buildings causes suffocation and death; that elevators sometimes refuse to work; that outside fire escapes are more serviceable as a means of entrance for firemen than as a means of escape for occupants; that the protection of the best fire department in the world is sometimes inadequate; and that persons of the highest intelligence often lose their presence of mind and become helpless victims of unreasoning panic. All of the above-mentioned precautions are good, but they do not go far enough. The people of the state of New York are no longer satisfied with measures that are merely good in matters involving life and death. They insist, within reasonable limits, upon the best measures of precaution that human and legislative foresight can devise, to that end that a recurrence of such tragedies as the Triangle Shirtwaist Company fire may in the future be prevented."

cases are those where the owner is prohibited from using his property. We are dealing here with a less drastic measure. But in no case does the owner of property acquire immunity against exercise of the police power because he constructed it in full compliance with the existing laws."

Professor Brannigan has summarized the applicable rules as follows:

1. There is no constitutional barrier against applying new codes to existing buildings;
2. Retrospective codes are subject to the due process clause of the Constitution;
3. A legislature's [or City Council's] judgment on the policy decision to insist on fire safety receives absolute deference from the court;
4. A legislature's [or City Council's] judgment on the existence of the hazard and the reasonableness of the means of control will be upheld if there is any reasonable evidence in support of them;
5. The balancing test required for due process is that (a) the hazard must be real, and (b) there must be at least some evidence that the means to control it reduces the hazard;
6. There being no constitutional requirement that the repairs be within the economic or technical capability of the owner, the owner can be required to demolish the building for public safety;
7. The weighing of a reasonable relationship between the means and ends is a question of law for a court and not a question of fact for a jury, although a jury can be used to determine the facts upon which the constitutional determination is based."[17]

In summary, if a regulation appears to be reasonably calculated to correct the evil at which it was aimed, and if the regulation does not impose too great a burden on private rights, it is clear that the courts will uphold the law and, in effect, sanction its application retrospectively.

From a practical standpoint, the best method of correcting fire hazards that involve the expenditure of large sums of money is to put aside the question of whether or not the measures can be required validly under existing laws. Instead, "sell" the person on the importance of installing the recommended safeguards to protect himself and his investment. The legal technicality of retroactivity is less apt to be brought into question if the law itself is not quoted to the person who is being "sold" on the measures. Rather, stress the underlying principles of fire protection, such as the subdividion of large areas, reduced concentration of high values in one area, and protection of vertical openings. Convince the person of the benefits which will accrue from complying with recommendations, such as increased life safety to employees,

---

[17] Brannigan, *op, cit.*, p. 448

insurance premium reductions, and lessened risk of civil liability. With this approach, the installation of fire protective measures often can be obtained quickly.

Theories of Liability

Owner/Operator Liability for Negligence

Sources of civil liability with respect to fire are found both in the common law and in statutes. Common law liability exists for injuries resulting from dangerous conditions of which the owner or occupant of the premises is aware. The general rule is stated in corpus juris (a body of law) as follows: "Notwithstanding a fire may be accidental in the sense that it is not intentionally kindled by him, one may be liable for injury occasioned by it where it is due to his negligence, or has failed to use ordinary care and skill to extinguish it, or to provide adequate means for so doing."[18]

The important question then becomes, what is the measure of care that is required? With respect to the occupant of property, the majority of jurisdictions hold that the duty of care varies with the relationship the occupier bears toward the injured party. (Yalowizer v. Husky Oil Co.)[19] In those jurisdictions, the duty owed business invitees is greater than that owed mere licensees or trespassers.

For example, suppose a fire breaks out in a hotel. There would be a high degree of care owed by the management to see that a paying guest is promptly notified in time to make any possible escape; that the route of egress has been posted in his room; that the halls are light enough so the guest can find the means of egress; and that access to the outside is unobstructed by locked doors, furniture, etc. To the individual who is not a paying guest of the hotel, but merely permitted, as a humanitarian gesture, to sleep in a chair in the lobby, only the duty owed a licensee would be required—a duty that does not extend to taking elaborate steps to protect the individual but involves merely warning him of any hidden dangers and not actively increasing any hazards already present. In this example, such a duty undoubtedly would extend to notifying the person of the fact that there was a fire and maintaining any safeguards ordinarily required by law, but would not obligate the management to see that he got out quickly or to underwrite the safety of his belongings. In the case of a trespasser, the duty

[18] 45 *Corpus Juris* 852

[19] The following states follow this rule: Alabama, Arizona, Delaware, Georgia, Idaho, Illinois, Indiana, Iowa, Kansas, Maryland, Mississippi, Missouri, Montana, Nebraska, New Jersey, North Carolina, Ohio, Oklahoma, Oregon, Pennsylvania, Texas, Utah, Washington, and Wyoming.

would extend to notifying him of the danger only if he was known to be on the premises.[20]

The minority of jurisdictions have abrogated all or some portion of the status scheme in favor of a uniform standard of care, specifically whether the owner in the "... management of his property has acted as a reasonable man in view of the probability of injury to others, and although the plaintiff's status as a trespasser, licensee, or invitee may in the light of the facts giving rise to such status have some bearing on the question of liability, the status is not determinative [on the issue of duty owed]." (*Rowland v. Christian*)[21] In our example, these jurisdictions would characterize the hotel owner's duty as a reasonable duty of care with respect to each individual—the paying guest, the gratuitous visitor, and the trespasser.

A smaller minority of states employ the uniform standard of care for plaintiffs whose status could be characterized as invitees or licensees. These states retain the status designation, and consequently the defined limited duty, for trespassers.[22] In our hotel example, the paying guest and gratuitous visitor would be owed a reasonable duty of care under the circumstances. The duty owed the trespasser would be notification of the fire should his presence be known.

Notice of a Hazardous Condition

Suppose a person were sprayed with atomized oil under high pressure issuing from a broken hydraulic line connected to a plastic molding machine; and suppose that, as so often happens, the spray breaks into flame and burns the person. Would the owner of the machine be liable to the injured person for the damages suffered? A person not trained in the law most likely would give a snap judgment without getting all the necessary facts. First, determine the relationship of the injured person to the management of the company to find out what duty, if any, was owed the injured party. Second, it is necessary to know if the injured person were in control of the machine and in any way negligent. For example, did he have any idea that the line was weak and about to break, and did he then fail to do what an ordinarily prudent person would do—replace the line? If he could have learned by routine inspection that the line was defective, was he obligated to perform such inspections but failed to do so, or did he fail to inspect with the frequency required by his employer or typically done by workers of comparable status and experience? If so, it would be considered negligence even if he did not actually know that the line was defective, for he should have known it. When a person, by using ordinary

---

[20] See 22 *American Law Reports* 4th 294, "Modern Status of Rules Conditioning Landowner's Liability Upon Status of Injured Party as Invitee, Licensee or Trespasser."

[21] Jurisdictions following this rule include: Alaska, California, Colorado, District of Columbia, Hawaii, Louisiana, New Hampshire, New York, and Rhode Island.

[22] These jurisdictions include: Florida, Maine, Massachusetts, Minnesota, North Dakota, and Wisconsin.

care, could find out something if he took the trouble, he is said to have "constructive notice" of the condition.

Assume that a fire, OSHA, or insurance inspector had been through the plant and warned the management that use of that particular type of machine was unsafe unless a nonflammable oil were substituted for the flammable type. After describing the reasons for the required substitution, the inspector provided examples of damages resulting from hydraulic line failures involving flammable hydraulic fluids. Under the circumstances, the management would have had notice. Irrespective of whether or not it made its own regular inspection, management would be under notice of the danger simply by the oral or verbal report of the visiting inspector. The fact that the law may not actually have required the changes recommended by the inspector is not material to the issue of whether or not there was notice of the hazard. If after such notice, whether actual or constructive, a reasonable time had elapsed in which corrective action could have been taken, but was not, then a jury might find that the management had been negligent.

As pointed out previously, the first inquiry would be to determine whether or not the person injured was owed a duty of care, and to see what degree of care would be required under the circumstances. If the person were an employee, there is a duty to provide reasonably safe equipment and premises. Of course, if the employee had seen the lines break on previous occasions and knew the dangers of the work, but voluntarily assumed that risk, then he may be precluded from a negligence recovery based upon the "Assumption of the Risk Doctrine." He still would be entitled to recover under the typical workers' compensation statute in effect in most states.

If the person injured were a salesman, customer, or other person who might be considered to be on the premises on a matter of mutual interest, he would be considered a business invitee. As such he clearly would be entitled to protection against such negligence provided that it was reasonable for him to go into the shop area to transact his business. If he had not been allowed to go into the shop and had been advised of the restriction and possible danger, then he would be owed a lesser duty of care.

While the duty to make inspections to discover defective machinery that might give rise to injuries extends to invitees, it does not extend to licensees or trespassers. If the licensee who cuts through the premises for his own convenience is warned of moving cranes or any dangers which are not obvious, and the company exercises ordinary care, that is sufficient; the licensee takes the premises as he finds them. And if he were the victim of the sprayed flaming oil, it is extremely doubtful that a jury would permit him to recover for his injuries. This would be true unless the fact that the line was about to burst had been brought to the attention of management who did nothing about it or failed to warn him of the danger.

The trespasser is in an even poorer position to claim that a duty of care owed him has been violated. The duty owed a trespasser is only to warn of hidden traps and the like, after his presence is known. If a man felling trees on his property saw a trespasser crossing his land in an area where the trespasser might get hurt, he would be required to warn the trespasser of the danger. Where a person stays in a zone of danger voluntarily (and not as part of his duty), any injury suffered as a result would be considered due to his own negligence. This is commonly referred to as "contributory negligence," the existence of which is a bar to recovery for damages.

Strict Liability of Property Owner/Operator

"Strict liability" is the concept whereby liability for a plaintiff's injuries is assessed on the basis that the defendant has chosen to undertake some specific act or activity, rather than on the failure of the defendant to exercise the required standard of care. In contrast to negligent activity, where the defendant is held "at fault," strict liability holds a defendant liable based upon social policy; despite no moral wrongdoing, it is better to shift the loss to the defendant for the greater good of society.

In the area of premises liability, the rules of strict liability apply to landowners when they undertake "abnormally dangerous activities." For example, suppose a person in the fumigation business were to store cylinders of methyl bromide or hydrocyanic acid on his premises. These are known to be extremely poisonous chemicals, with vapors more insidious than most poison gases because, being odorless, they can penetrate even brick walls without giving any noticeable warning of their presence. If these chemical containers were to leak, it is very likely that individuals in adjacent structures could be killed without warning. Were the fumigator held only to a duty of reasonable care, he potentially could defend his actions by showing that the means he employed for storage and inspection were reasonable. Such a defense would be truly unjust.

However, the judge of this case more likely would find that keeping of these poisonous gases was such an inherently dangerous activity that strict or absolute liability should be imposed. There would be no necessity of showing that the owners failed to use reasonable care. Unusually dangerous activities have included collecting water in quantity in a dangerous place, or allowing it to percolate; storing explosives or flammable liquids in quantity in the midst of a city; blasting; pile driving; crop dusting; fumigation with cyanide gas; drilling oil wells in populated areas; allowing factories to emit noxious smoke; and permitting dangerous party walls.[23]

Strict Liability of Product Manufacturer/Distributor

The concept of strict liability also is applied to the area of injuries resulting from defective products. The rules provide that the product manufacturer

---

[23] Prosser, William L., *Handbook of the Law of Torts*, West Pub. Co., 4th Edition, 1971, pp. 509-510

and all others in the product's distribution chain will be held strictly liable, in tort, for any defect in the product which was the proximate cause of a plaintiff's injury. *Restatement (Second) of Torts*, Section 402A, states the applicable rule:

> "1. One who sells any product in a defective condition unreasonably dangerous to the user or consumer or to his property is subject to liability for physical harm thereby caused to his property, if
> A. the seller is engaged in the business of selling such a product, and
> B. it is expected to and does reach the user or consumer without substantial change in the condition in which it is sold.
>
> "2. The rule stated in Subsection 1. applies although
> A. the seller has exercised all possible care in the preparation and sale of his product, and
> B. the user or consumer has not bought the product from or entered into any contractual relation with the seller."

The rules of product liability have been applied to fire trucks, ladders, hose, fire extinguishing systems, even smoke detectors. However, in *Smith v. F.W.D. Corp.*, the city, which compiled plans and developed specifications for fire trucks, was not liable to a fire fighter under a strict liability theory, for the city did not design the vehicle nor serve any role in the product distribution chain other than as the ultimate consumer.

It is important to note that if a fire department regularly engages in the sale of fire extinguishers, smoke detectors, or any other fire protection device to promote local fire safety and support the department, it could be construed to be "in the business," as part of the distribution chain, and could be held strictly liable for any injuries resulting from a product's defective condition. However, were the department only to endorse the general purchase of fire safety devices or provide names of "approved" or "listed" devices, it would not be exposing itself to potential strict liability.

Res Ipsa Loquitur

In other situations, to obtain a favorable judgment it may not be necessary to prove that a defendant's conduct breached some duty of care. These are situations where the doctrine of *res ipsa loquitur* applies. The literal translation of this Latin phrase is "Let the thing speak for itself." The doctrine serves to raise a presumption or permit an inference of negligence on the part of the defendant.

Because everyone is presumed to exercise ordinary care in his activities, the burden of proof usually is on the person bringing the law suit to establish that the defendant has been negligent—that the defendant has not acted as an ordinary prudent person would under the circumstances. This presumption that a person uses ordinary care is, however, rebuttable. When the doctrine of *res ipsa loquitur* is applied, instead of bringing in evidence to prove the defendant's negligence, the plaintiff establishes what is called a *prima facie* case. For example, facts surrounding the fire show that all the factors necessary for application of *res ipsa* are present. The rules for application of the doctrine vary by state common law and statute, but they principally require:

1. All of the factors and instrumentalities causing the injury were under the exclusive control of the defendant.
2. The accident was of the sort that would not occur but for the negligence of someone.
3. Plaintiff is not contributorily negligent.
4. The evidence as to the cause is more easily accessible to the defendant than to the plaintiff.[24]

The *prima facie res ipsa* case is established when the plaintiff shows enough facts so a judgment could be rendered in his favor if no evidence is presented by the defendant in rebuttal. If the four factors listed above are present so that negligence can be inferred, the burden of going forward then is transferred to the defendant. He must prove that he was not negligent in order to avoid liability.

For example, suppose the night clerk in the hotel example discovers a fire in a trash pile around midnight. Instead of calling the fire department immediately, he starts ringing all the rooms. During the 20 minutes that elapse before the fire department finally is alerted, a man suffocates from the heat and fire gases in his room. Further, management never had instructed the clerk on how to sound the alarm gong nor on the general procedure to follow in case of fire. From these facts alone, it is possible under *res ipsa loquitur* to infer that the death would not have occurred but for management's negligence. (*Parker v. Kirkwood*)

In *Aetna Insurance Co. v. 3 Oaks Wrecking and Lumber Co.*, fire occurred in a vacant, unsecured building in the process of demolition. Sited in a "skid row" section of the city, the building was without electrical, gas, or other utility service. It was used by vagrants, alcoholics, and hoboes as a place to smoke, cook, and sleep. The defendant was under a code obligation to secure the premises against unauthorized entry and to demolish the structure, one story at a time. Despite the code requirements, the defendant had created numerous vertical openings which permitted rapid development of the fire and its spread to adjacent properties.

The court noted "that in order for the doctrine to be applicable it was necessary that the accident be of a kind which does not ordinarily occur in

---

[24] See 21 *American Law Reports* 4th 929, "*Res Ipsa Loquitur* as to Cause of or Liability for Real-Property Fires."

the absence of someone's negligence and that the instrumentality which causes the injury must be in the exclusive control of the defendant.... Only when the fact of the fire plus the surrounding circumstances give rise to an inference of negligence is *res ipsa loquitur* applicable.... In the case at bar, the nature of the neighborhood, the fact that the doors were open, copper had been stolen, the problems with vandals in the building next door, the fact that there was no source of energy or power, and the fact that the event occurred on a clear, dry day were circumstances that do give rise to an inference of negligence."

In addition, it has been held that a plaintiff has made a *prima facie* case under *res ipsa loquitur* in the following instances:

1. A dehydrator tank exploded during welding operations. There was no water in the tank which had held flammable liquids. The defendant had told the welder that the tank was prepared for welding operations by washing and filling it with water, and by demonstrating that the tank was filled.
2. Action taken by a store owner against a neighboring store where evidence showed fire occurred in the neighbor's stockroom—an area without utility service where stockmen customarily smoked.
3. Another action was against a warehouseman, whose employee was seen using an open flame lamp near the point of origin of a fire.
4. Suit was brought against a hotel owner by a plaintiff who lost his son in the panic of a fire. The plaintiff threw himself through a second-story window which would not open and fell to the ground. The hotel's halls were inadequately lighted and there was no alarm system.

Liability for Spread of Fire

A person is liable for injuries resulting from negligence in not taking proper precautions to prevent fires or their spread. This common law rule is adopted by statute in many states and covers both criminal and civil liability. The rule applies to any person who, himself or through another person, willfully, negligently, or in violation of law sets fire to, allows a fire to be set to, or allows a fire he kindled or attended to escape to the property (public or private) of another without exercising due diligence to control such fire. This person is liable to the owner of such property for the damages caused by the fire. Also, the expense of fighting such a fire is a charge against the person responsible and may be collected as if the expense were part of a written contract.

Many cases hold that the owner of a building is liable to the owners of surrounding buildings for fire in his building caused by his negligence.

In an exemplary incident, a tenant left on a pressing iron in her small apartment's kitchen. The iron burned a hole through the ironing board without setting the room afire, but it did smoke up the kitchen paint considerably. The tenant offered to pay the landlord the cost of repainting the kitchen, estimated by a painting contractor at $100. However, the landlord chose to collect the $500 offered by the insurance adjuster and refused the tenant's offer. The insurance company then collected the $500 from the protesting tenant.

The duty of care to limit outbreak or spread of fire also can apply to government. In *Lorschbough v. Twp. of Buzzle*, a fire of unknown origin began in the township dump, was quickly contained, but was left to smolder for many days. Ten days after the initial outbreak, a series of strong wind gusts disturbed the earth covering the dump fill and swept burning straw and garbage into nearby weeds and timber. A forest fire ensued, destroying various real and personal property of neighbors. In holding the township liable for spreading the fire, the court took specific note of the township's failure to comply with standards prescribed by the Pollution Control Agency with respect to controlling hazards at sanitary fill sites.

Negligence Per Se

Violation of a law or ordinance may amount to *negligence per se*; that is, negligence even in the absence of proof that would show failure to exercise ordinary care. (*Per se* means "as such" or "in and of itself.") The reason is that the regulation itself establishes the standard of care which a person should follow. If he fails to comply, the law has the effect of proving failure to exercise due care.

For example, an incinerator located too close to the eaves of a garage set the shingles on fire. The burning shingles ignited the roofs of other garages along the alley. The owner of the garage where the fire started was reimbursed by his insurance company for the damage to his garage, but the insurance companies which paid the claims of the owners of the other garages sued the owner of the incinerator. Judgment was against him on the grounds that the law required the stack to be at least five feet away from the wooden garage and this one was only two feet away. Only a week before, a fire inspector who looked at the installation had tacitly approved it by not calling any violation to the attention of the owner. This fact, however, did not constitute a good defense.

As with most property insurance policies without personal liabiity provisions, the incinerator owner's insurance did not cover damage to property other than his own. The fire inspector perhaps thought he was doing the owner a kindness in not requiring him to move the incinerator. If it had been only a matter of a few inches one way or another, it would have made little difference because such a location probably would not have caused the fire. As it was, the difference in location was sufficient to create the hazard. Under the typical ordinance found in many communities, the owner thus was guilty of a misdemeanor or infraction in "permitting" the condition to exist in violation of the law. Usually, where it is a misdemeanor to create a hazard, it also is a misdemeanor to aid, suffer, abet, or permit the continuance of the hazard. Each day the hazard continues is a separate offense; this keeps the statute of limitations from tolling and thus preventing prosecution of the person creating the hazard.

Some fire escape cases also are examples of how liability can be imposed where violation of an ordinance or order of the fire department results in injury. In one case, *Mars v. Whistler*, it was held "negligence *per se*" to maintain a fire escape with a hatchway larger than was allowed by law. Both the landlord and tenant were held liable as a result of the plaintiff's fall through the opening. Other cases (*Roxas v. Cogna and Hoopes v. Creighton*) held the owners of property responsible because no fire escapes had been provided as required by law.

In a case growing out of a restaurant fire, an inspector apparently had overlooked an illegal grease duct connection between the range hood and the general ventilation system for the building. The system did not have the insulation or clearances from combustible framework required for regular grease ducts. The duct had been installed many years before the defendant took over the premises, and the fire marshal had been in and out of the place many times without ever noticing the illegal connection. Nevertheless, the court sustained the jury's decision against the restaurant for damages to the property of others caused by the extending grease duct fire. Hence, this case illustrates how even an unintentional violation of the law may result in liability for violation of a code or ordinance requirement.

Code Enforcement

Local Ordinances

A principal method which a fire prevention bureau employs to promote fire safety, and perhaps the most common method, is the adoption and enforcement of building and fire prevention codes. This involves the filing of complaints and imposition of fines or other penalties for code violations. Although communities have the power to promulgate their own codes, the usual practice is to adopt one of the model codes, by reference, through promulgation and passage of an ordinance. Therefore, fundamental understanding of the requisite elements of an adoption ordinance is important.

A "municipal ordinance" sometimes is referred to as a by-law or local law—a regulation of a general, permanent nature or a legislative act of the municipal governing body.[25] By contrast, a "resolution" is a temporary action; it usually is employed to state the sense or feelings of the city government on an issue, or to direct performance of some administrative act.

Because an ordinance carries the force of law, its enactment must comply with all the statutory formalities for its promulgation. Requisites for a valid

---

[25] Rhyne, *op. cit.*

ordinance, in addition to all the essential requisites for its passage, generally are summarized as follows: 1. The ordinance must have a clear, reasonable, and substantial relation to the public health, safety, morals, or welfare; 2. It must be reasonably appropriate for the police power objective it is designed to achieve; 3. It must be reasonably definite, clear, and certain—that is, definite enough to serve as a guide to those to whom it applies; 4. It must not be unreasonably discriminatory and must apply uniformly to all members of the classes affected; and 5. It must conform to provisions of both the Federal and state constitutions, including the common law where it is the law of the state.[26]

"Quite often courts are reluctant to review legislative decisions. A rule of reasonableness coupled with wisdom and propriety is employed by the courts in reviewing legislative decisions. If this reasonable rule is debatable, the courts have historically declined to become involved. Thus, many debatable legislative decisions are automatically upheld. However, if the rule of reasonableness has been abandoned, the courts will review the premises upon which the legislative decision was based."[27]

An ordinance may not serve to transfer the discretionary, legislative power from a council or legislature to an administrative agency or officer. It may not provide for arbitrary discretion of, or fail to provide reasonable standards and guidelines for, administration and enforcement officials. In general, all such ordinances must be confined to the affairs of the community and be applicable only within its corporate limits.

No particular formality in wording or passage is essential to the validity of a municipal ordinance, unless specific requirements appear in the city's charter or state statutes. However, any ordinance must indicate unmistakably the intent by government to assert its will. Such enactment clauses should be drafted carefully, for they frequently are reviewed by the courts in an effort to determine the city's intent at the time a specific ordinance was passed. Unfortunately, ordinance provisions often are truly ambiguous.[28]

---

[26] Rhyne, *op. cit.*, pp. 452-3

[27] "Recent Developments on Ordinances and Administrative Regulations," *Urban Lawyer*, Vol. 14, No. 4, Fall 1982, pp. 855-6

[28] The model codes typically include a recommended enactment clause for utilization during adoption of the code by reference. For example, the Uniform Building Code, 1985 Edition, published by the International Conference of Building Officials, provides the following recommended enactment clause: "An ordinance of the (jurisdiction) adopting the 1985 edition of the Uniform Building Code and the 1985 edition of the Uniform Building Code Standards regulating the erection, construction, enlargement, alteration, repair, moving, removal, demolition, conversion, occupancy, equipment, use, height, area, and maintenance of all buildings or structures in the (jurisdiction); providing for the issuance of permits and collection of fees therefor; providing for penalties for the violation thereof, repealing Ordinance No. (old code) of the (Jurisdiction) and all other ordinances and parts of the ordinances in conflict therewith."

Careful consideration also must be given to the content of proposed ordinances. The draftsman should have sufficient farsightedness to consider the entire field instead of the particular problem facing him at the moment. Many times, a new ordinance is written whenever those on the books are not adequate to handle the present situation; this only results in a disjointed compilation of "spur-of-the-moment" requirements. Ordinances relative to fire prevention should be revised as a body, rather than as individual units, if the proper relationship among the various subdivisions is to be maintained.[29]

Reasonable underlying principles of fire safety should be the basis of every applicable law enacted. Thus, in applying these ordinances it becomes unnecessary to quote the exact wording of the law as authority for making a particular recommendation. Also, as a code cannot conceivably cover all situations, reference to the underlying principles employed in its promulgation will serve to guide the code enforcing authority in the resolution of problems. In addition, should the code requirement be challenged, any appeals body or the court will give greater deference to the code authority where the underlying basis for the code requirement can be stated.

Another consideration in drafting municipal ordinances is how to draw up local regulations which will not duplicate nor conflict with state laws on the same subject. This topic, called "conflict of jurisdiction," will be covered later in the section bearing that title.

A more perplexing question is whether or not a municipality can impose penalties for acts which also are declared to be crimes under the laws of the state. Specifically, can prosecution under the ordinance be pleaded as a bar to a prosecution under the statute? For example, can a person who has been arrested for selling fireworks in violation of a city ordinance be imprisoned for violating a state law to the same effect?

Before 1850, it generally was assumed that, in the absence of express authorization, an ordinance was void if it attempted to punish acts already criminal under the laws of the state. The majority view was that when both an ordinance and a state law prescribed a penalty for a given act, a conviction or acquittal under one was a bar to prosecution under the other.

---

[29] To preclude conflicting overlap among local government laws, many are compiled or "codified" as discussed previously for state laws. "Codification has been defined as the process of collecting and arranging the ordinances of a specific local government unit into a complete set of positive law, logically arranged and promulgated by the legislative authority—that is, a code. The codification of local laws thus involves not only a compilation of existing laws, with systematic arrangement, cross-referencing, and indexing, but also a legal review of the content of such laws to identify and resolve conflicts among the various regulations and amendments to those regulations, to clarify ambiguities, and generally to make whatever changes, additions, or deletions are needed to produce a complete and efficient system for the regulation of those subject areas with which the law deals." "Recent Developments," *supra*, p. 858

Now, however, it is well settled that the same act may constitute an offense against both the state and the municipality and may be punished by either or both. Various reasons are advanced for this rule, but an increasingly popular one is based on the analogy to the similar rule sustaining successive state and Federal prosecutions. As each is an independent sovereign, with independent rights to promulgate laws for its own protection, each has independent authority to assure compliance with its laws. However, in sustaining the rule, courts are careful to point out (*In Re Mingo*) "that we only hold that there is a conflict where the ordinance and general laws [of the state] punish precisely the same acts. We do not wish to be understood as holding that the sections of the ordinance which make criminal other acts not punishable under the general laws [of the state] are void because the legislature has seen fit to legislate upon the same subject."

The courts have been very liberal in interpreting that city ordinances do not conflict with state regulations. For example, in *Ex Parte Boswell*, the court held that a statute prohibiting gambling does not prevent a municipal corporation from making it a misdemeanor to visit a place for the purpose of gambling. The general rule on concurrent legislation can be said to be that "the only way the legislature can inhibit local legislative bodies from enacting rules and policing regulations is by the state occupying the same legislative field so completely that legislation on the subject by local legislative bodies will necessarily be inconsistent with the state act." The legislative intent and comprehensiveness must clearly preempt the area of the subject matter from local regulation in order to prohibit promulgation of additional local requirements. (*In Re Iverson*)[30]

Safety Standards

It is not necessary that there be a violation of an ordinance or statute to establish liability for the occurrence or spread of fire. This rule was aptly stated by the court in *Luxen v. Holiday Inns* in the context of a slip and fall case: "While defendant may well have been in compliance with the applicable building and safety code provisions, such compliance does not preclude a determination that, under the circumstances, defendant was nevertheless negligent." As Prosser notes, compliance with a statute does not necessarily mean that due care was used.[31] Thus, again from *Luxen v. Holiday Inns* "where specific circumstances present situations beyond those which the statute was designed to meet, a plaintiff may prove that the defendant was negligent in not taking extra measures."

National standards, such as those promulgated by the National Fire Protection Association, can have an important bearing on the question of negli-

---

[30] See also: "Recent Developments," *supra*, 901.

[31] Prosser, *op. cit.*, Section 36, p. 203

gence in providing adequate fire protection. Specifically, where a standard has been promulgated before the date of an accident, this fact will serve as illustrative evidence of safety practices generally available and potentially prevailing in a given community. Such standards may demonstrate the present thinking in the field of safety, and a recognition and codification of customs and practices.[32]

A case in point is the action brought against a marina to recover damages by fire to a yacht in winter storage. The court held that the marina, "as bailee of the yacht for winter storage, took none of the precautions required by the Fire Protection Standard for Marinas and Boatyards (NFPA 303) published by the National Fire Protection Association, employed no watchman, and provided no water source .... The marina was negligent and its negligence was the proximate cause of the spread of fire [from point of origin] to the nearby yacht, notwithstanding that neither party had information revealing the origin of the fire." (*Firemen's Fund v. Captain Fowler's Marina*)

In a personal injury action (*Frazier v. Continental Oil*) arising out of a flash fire at a service station, there was evidence that the vent pipes running from gasoline tanks violated standards set forth by the NFPA. The evidence was ruled admissible where it was relevant to determination of the appropriate installation practice.

Remember, however, that the fact of violating a national safety standard is not a conclusive determination of negligence. Rather, the court may permit admission of a standard into evidence merely in order to assist the jury in its deliberations.

Other Methods of Promoting Fire Safety

In addition to filing complaints for violations of ordinances, abating nuisances, and stressing personal liability, other legal procedures often are employed to promote fire safety.

Some departments require the posting of bonds, which may be forfeited under certain conditions; for example, upon the breach of lawful conduct in the operation of a business granted a permit; or upon the failure of a sprinkler installer to have his work comply with installation standards. The proceeds upon such a breach (the value of the bonds forfeited) can be used to defray the expense of putting the property in a code-complying condition. Such is the practice in Los Angeles with respect to oil wells, refineries, and natural gas plants, for which a bond is required with the application for a permit for maintenance or operation.

---

[32] See: Comment, "Admissibility of Safety Codes, Rules and Standards in Negligence Cases," 37 *Tennessee Law Review* 581 (1970).

See also: 58 *American Law Reports* 3d 148, Annotation, "Admissibility in Evidence, On Issue of Negligence, of Codes or Standards of Safety Issued or Sponsored by Governmental Body or by Voluntary Association."

Licensing, with the possibility of revocation of the license, is another effective control in the field of fire prevention. The fire department could use the licensing as a device to regulate those businesses which affect the public safety. License fees could be used to cover the fire department's costs for performance of inspections necessary to assure the licensee's fire code compliance. Note that such fees should amount to no more than is necessary to defray inspection and supporting clerical expense. The courts are not exacting in the determination of the cost of regulation and supervision, yet any amount disproportionate to the expense of issuing the license and regulating the business would be considered unreasonable and invalid as a matter of law.

Conflict of Jurisdiction

Before leaving the subject of fire prevention laws and their enforcement, it might be well to consider the problems that can arise with respect to multiple and conflicting requirements of various political agencies regarding a given subject. It already has been noted that enforcement of municipal ordinances which do not conflict with, nor regulate in exactly the same manner as, state laws usually are valid. The same reasoning and analysis can be applied to Federal regulation. Careful review of, and possibly reference to, the regulations of the state or Federal agency might be required to preclude a finding of conflict. Also, examination of the enacting clauses of the law or ordinance might clarify the question of whether the Federal, county, or state government intends to preempt regulation in the particular field.

Federal Regulation

"Occasionally, a city may find that some area of law-making has been preempted [or 'occupied'] by the Federal government so as to make any local regulation impossible, just as such preemption sometimes occurs through state action. Of course, Congress in legislating for the Federal government is limited to areas of constitutionally recognized power—but some such areas, as that covered by the Commerce Clause [of the U.S. Constitution], have proven quite broad indeed. Assuming that the requisite power is present, a Congressional intent to occupy a field of possible regulation will still not be assumed, but will be found if any one of three conditions exists: 1. an express legislative provision of preemption, 2. a scheme of Federal regulation so pervasive as to lead to the reasonable inference that Congress left no room for additional regulation, or 3. Congressional legislation touching a field in which Federal interest is so dominant that the Federal system is assumed to preclude state or local laws on the subject. In the first instance, express preemption has occurred, and in the second and third instances, implied preemption."[33]

---

[33] Reynolds, *op. cit.*, pp. 131-2

Regulation of the Transport of Hazardous Materials as an Example of Conflict

The Federal power to regulate transportation of hazardous materials is derived from the Commerce Clause of the Constitution.[34] On purely local matters, not involving or affecting interstate or foreign commerce, the Federal government has no regulatory power, no matter how dangerous the transported material may be; in such instances, the police power is vested solely in the state and local governmental bodies. (*Boston and Maine Rail Road v. Armburg; Brolan v U.S.*) However, where there is any potential for impact upon interstate commerce, the authority of the Federal government to regulate generally will prevail.

The question arises as to how far the Federal government can go in regulating the shipment and transportation of dangerous substances, such as ammonium nitrate, hydrocyanic acid, liquid oxygen, or fluorine. Under the Commerce Clause of the Constitution, the power of Congress over interstate and foreign commerce is practically absolute. As stated in *McCulloch v. Maryland*, "the sound construction of the Constitution must allow to the national Legislature that discretion, with respect to the means by which the power it confers are to be carried into execution, which will enable that body to perform the high duties assigned to it, in the manner most beneficial to the people. Let the end be legitimate, let it be within the scope of the Constitution, and all means which are appropriate, which are plainly adapted to that end, which are not prohibited, but consistent with the letter and spirit of the Constitution, are constitutional."

Thus, a safety regulation pertaining to hazardous materials transport, promulgated by a Federal agency under an Act of Congress, is enforceable provided it is not arbitrary nor capricious. The regulation must be consistent with the hazards involved; if experience has shown that the transportation involves serious dangers, the regulation may be comprehensive. It is evident that transportation of dangerous chemicals must be considered as falling within the class of serious hazards. As an example of where possible conflicts can arise, consider some issues involved with, and the current Federal regulations governing, transportion of hazardous materials.

The purpose of the Hazardous Materials Transportation Act is stated to be to "improve the regulatory and enforcement authority of the Secretary of Transportation to protect the Nation adequately against the risks to life and property which are inherent in the transportation of hazardous materials in commerce."[35] Under such authority, the Secretary effectively has authority to

---

[34] Congress shall have the power "to regulate commerce with foreign Nations, and among the several States, and with the Indian Tribes...." Article I, Section 8, United States Constitution

[35] 49 U.S. Code Supplement Section 1801 et seq. (Jan 3, 1975, P.L. 93-633, Title I, Section 102, 88 Stat. 2156)

promulgate regulations affecting the packaging of shipments of hazardous materials; packaging markings (to show content) and labeling (to show hazard); vehicle placarding (to show hazard); handling procedures, such as loading and unloading requirements; routing; care of vehicle and its cargo during transportation; and preparation and use of shipping papers to show identity, hazard class, and amount of each hazardous material being shipped. Such regulations, having the force of law, apply to persons who offer hazardous materials for transportation in commerce (shippers), those who transport the materials in commerce (carriers), and those who manufacture and retest the packaging and other containers intended for use in the transportation of materials in commerce.[36]

State or local regulations affecting the transportation of hazardous materials have been virtually preempted by Federal regulations in the areas of: hazardous class definitions, hazard warning systems, shipping paper requirements, and container requirements. Requirements for local permits have been held invalid where the permit documentation duplicates Federal requirements and would result in unnecessary transportation delay. In view of the legislation's goal of providing uniformity in transportation regulations, state or local requirements limiting the time, date, or place of transportation generally have been held preempted. Exceptions are made where a clearly defined local need is covered by local requirements that agree with Federal intent.

For example, in *City of New York v. Ritter Transportation*, the court reviewed the regulations prohibiting transportation of bulk flammable gases through New York City except in accordance with routes and times approved by the Fire Commissioner. The effect of the regulations was to ban interstate transportation of hazardous compressed gases in tank truck quantities in New York City, except for a single, highly circuitous route for deliveries to Long Island. In sustaining the regulations, the court noted that the intent of the Federal regulations was to improve safety and to expedite the movement of hazardous materials travel—especially through use of interstate highways traversing limited population areas. Although use of the single permissible route to Long Island entailed travel almost completely around New York City, and raised costs by 35 percent, the regulations were upheld as improving safety in accordance with the Federal act and not being overly burdensome on interstate commerce.[37]

---

[36] See "Hazardous Materials: DOT Inconsistency Rulings," IR-7 through IR-15, 49 *Federal Register* 46632 (1984). The entire subject of Federal hazardous materials regulation is covered in detail at: Annotation, State or Local Regulation of Transportation of Hazardous Materials as Preempted by Hazardous Materials Transportation Act (49 USCS Section 1801 et seq.) 78 *American Law Reports* Fed. 289.

[37] See: Schwartz, Allen G. and Barry L., "New York City Defeats Challenge to Hazardous Materials Regulations," *Municipal Attorney*, Vol. 23, No. 1, Jan.-Feb. 1982.

However, in *National Tank Truck Carriers v. Burke*, state regulations prohibiting transportation by land of certain hazardous materials during workday commuting hours was held to be preempted by Federal regulations. State regulations were seen to undermine the intent of the Federal regulations to expedite hazardous materials transport. In addition, the court noted that the impact of such regulations would be to shift the burden of hazardous materials transport to adjacent states, contrary to Congress' intent to achieve national uniformity in transport requirements. Although hazardous materials transport through a community could be controlled, it could not be prohibited effectively.

State and Local Regulations

"It is generally recognized that local government may pass laws which go beyond a state statute [or Federal regulation] governing the same subject as long as the local law is not in direct conflict with the statute [or Federal requirement] and the legislature [or Congress] has not preempted the regulation of the field. The mere fact that the local law goes beyond the state statute [or Federal requirement], in requiring more, . . . is permissible as long as the two can co-exist . . . and the local law does not contravene the intent and purpose of the statute[or regulation]." (*Haddenham v. Board*)[38] The general rule is that where a local law merely enlarges upon the provisions of a state statute or Federal regulation by having stricter requirements, there is no conflict between the two unless the legislature or Congress has preempted regulation in the field.

Returning to the hazardous materials example, although local governments may not regulate the transport of hazardous materials, they can be expected to regulate the dangers created by the presence of such materials within their local jurisdictions. As has been shown, most regulations governing the packaging, labeling, and documentation of hazardous materials have been preempted by the Federal government. However, those regulations do not fully address all issues involving hazardous materials, such as their storage and dispensing. As Congressional authority is limited to regulation of hazardous materials in interstate commerce, issues involving their local use remain the province of state and local government.

For example, provisions of state and local fire prevention codes, and the model fire prevention codes frequently adopted by reference, typically address the subjects of where hazardous materials may be stored, storage amounts by individual occupancies, their separation from adjacent occupancies, construction of hazardous materials storage facilities, drainage, fire suppression systems, and hazardous area electrical systems, and security.[39]

---

[38] This case held that state law did not preempt county regulation or prohibition of the sale of fireworks.

[39] For example: NFPA 1, Fire Prevention Code, 1982 edition, Section 3-9; Uniform Fire Code, 1985 edition, Article 80

Regulation of hazardous materials is truly an amalgam of Federal, state, and local requirements, with some governmental entities possessing greater authority or exercising greater control than others.

Local Codes

The common practice of inspecting occupancies for building, fire code, and permit violations gives rise to some interesting questions. What about a fire inspector who discovers violations of building ordinances while he is inspecting for violations of fire ordinances? Should he refuse to approve the issuance of a permit by the fire department where he finds that building code violations make the building unsafe for the use intended? Notwithstanding the fact that enforcement of building ordinances may be the exclusive responsibility of the building department, the fire department should withhold approval of its inspection or permit until the matter has been referred to the building department for investigation. The fire department should not approve an unlawful condition, even though the condition is not a matter affecting fire safety. If the department does so, it may expose itself to liability.

The better practice for the fire inspector, in surveying the premises, would be to submit a complete written report of all conditions which he thinks should be brought to the attention of his superiors. This way, there is no necessity for an individual inspector to assume the responsibility of deciding whether or not the presence of certain conditions should prevent an applicant from obtaining a permit or passing an inspection. It should not be left to the inspector to approve or disapprove, but merely to investigate and report. Such a report should include a statement of all dangerous or defective conditions that come to the inspector's attention, so his superiors can determine the appropriate action.

It is a pleasant commentary on modern fire prevention to note the increasing tendency toward joint inspection tours by city building, health, and fire inspectors. This serves to facilitate communication among agencies as well as correction of discovered hazards in cases where the responsible department is in question. Needless to say, the effect on the owner is a positive one, for he is able to find out, at one time, whether he is in compliance with codes. Joint inspections also serve to limit interruptions in the performance of the owner's work.

Municipal and Inspector Liability for Enforcement or Nonenforcement of Statutes and Ordinances

Under common law, local governments have not been held liable for injuries resulting from enforcement or failure to enforce statutes or ordinances.[40]

[40] See *Restatement (Second) of Torts*, Section 315 (1964).

In regard to enforcement, this even has included enforcement of laws that were void, as in *Cheektowaga v. Buffalo*. "Absent a special relationship creating a municipal duty to exercise care for the benefit of a particular class of individuals, no liability may be imposed upon a municipality for failure to enforce a statute or regulations," as stated in *Sanchez v. Liberty*.[41] Liability can be found only where a municipal code shows a clear intent to identify and protect a particular, clearly defined class of people, according to *Halvorsen v. Dahl*. However, where statutory immunity applies to enforcement or failure to enforce a law, recovery will be impossible.[42]

See also *Grogan v. Commonwealth*. Court found no liability for failure to enforce building and fire code provisions, violation of which ultimately contributed to the disastrous Beverly Hills Supper Club fire.

The policy reasoning for holding no liability in the absence of a special relationship or explicit statutory obligation was enunciated in *Trianon Park Condominium Association v. City of Hialeah*. "The discretionary power to enforce compliance with the building code flows from the police power of the state. In that regard, this power is no different from the discretionary power exercised by the police officer on the street in enforcing a criminal statute, the discretionary power exercised by a prosecutor in deciding whether to prosecute, or the discretionary power exercised by a judge in making the determination as to whether to incarcerate a defendant or place him on probation. Statutes and regulations enacted under the police power to protect the public and enhance the public safety do not create duties owed by the government to citizens as individuals without the specific legislative intent to do so." Note that the employer-employee relationship, by itself, does not create a special relationship requiring the city to enforce ordinances for the protection of its employees. *Quinn v. Nadler Bros.* said the city owes no special duty to its fire fighters for its failure to enforce city ordinances designed to protect the general public from health, safety, and fire hazards.

Municipal and Inspector Liability for Negligence

A major area of fire department litigation has been in fire prevention services, specifically the performance or nonperformance of fire inspections, the care undertaken during the inspections, the nature of reporting and communicating the inspection results, and the follow-through on uncovered code

---

[41] In a wrongful death action based upon a fire in a residential occupancy lacking required fire escapes and adequate exits, the court of appeals held that the trial court erred in failing to dismiss a complaint against the village and its building inspector, where no special relationship existed to create a municipal duty to exercise care for the benefit of a particular class of individuals.

[42] For example. Kansas Statutes Annotated, Section 75-6104 (c), grants immunity for: "enforcement of or failure to enforce a law, whether valid or invalid, including, but not limited to, any statute, regulation, ordinance or resolution."

violations.[43] Historically, the fire prevention service was immune from critical examination; however, an expanding minority of states are holding the service liable for failure to exercise due care in the performance of its functions.[44]

In examining the issue of negligent building or fire prevention inspection, it is important to recall the elements that form the *prima facia* case of negligence.[45] Principally, it is on two of these elements—duty and causation—that the question of inspection liability turns. Specifically, did the inspector (city) actually owe a duty to the plaintiff, and did the inspector's (city's) negligent act bear a causal relationship to the plaintiff's injuries? Remember, all four elements of the negligence cause of action (duty, breach, causation, and damages) must be present to establish liability. However, as breach and damages generally are factual issues to be resolved by the jury hearing the action, the legal question of liability focuses upon whether or not the plaintiff actually was owed a duty and if the breach of that duty caused the plaintiff's injuries.

Under common law, there is no duty on the part of government to carry out fire or building inspections. Thus, government has no liability for its failure to perform such inspections. Fire inspection services have been considered a governmental function, owed only to the general public and not performed or owed to individual members of the community. The courts have held: "There can be no liability for a mere failure to be an instrument of good, i.e., for a mere failure to prevent harm caused by someone else, unless a statute pre-

---

[43] The term "inspection" as used here means investigation or examination for the purpose of determining whether any property other than property of a governmental entity complies with or violates any law or regulation of the governmental entity or constitutes a hazard to public health or safety.

[44] Sellers, Gary L., "State Tort Liability for Negligent Fire Inspection," *Columbia Journal of Law and Social Problems*, Vol. 13, 1977, pp. 303-350

"Municipal Liability for Negligent Building Inspection or Failure to Inspect," *Urban Lawyer*, Vol. 14, No. 4, Fall 1982, p. 687

[45] "The traditional formula for the elements necessary to such a cause of action may be stated briefly as follows:

A. A duty, or obligation, recognized by the law, requiring the actor to conform to a certain standard of conduct for the protection of others against unreasonable risks.

B. A failure on his part to conform to the standard required. These two elements to make up what the courts usually have called negligence; but the term quite frequently is applied to the second alone. Thus, it may be said that the defendant was negligent, but is not liable because he was under no duty to the plaintiff not to be.

C. A reasonably close connection between the conduct and the resulting injury. This is what is commonly known as 'legal cause' or 'proximate cause.'

D. Actual loss or damage resulting to the interests of another . . . ."

Prosser, *op. cit.*, p. 143

scribes such a duty."[46] This remains the majority rule, with exceptions; specifically, if a statute requires inspections, or if a judicially recognized "special relationship" exists between the plaintiff and inspector. (*Cracraft v. City of St. Louis Park*)[47]

However, a minority of states have expanded the inspection obligation by holding that duty could be based upon the city's permissive authority to undertake inspections. In those jurisdictions, the inspector will be under a duty to exercise reasonable care throughout all inspection activities, including decisions to inspect, performance of inspections, recording and notification of results, and enforcement of code-required corrective actions despite the lack of common law or statutory duty to perform in the first place. The inspector's duty will be based solely upon his authority to perform the service. This was affirmed in *Stewart v. Schmeider*, *Adams v. State*, and *Coffey v. City of Milwaukee.*

In the majority of states, the statutory duty to perform an inspection must be explicit, stated *Gordon v. Holt.* The courts will not imply a duty nor expand the bounds of such duty by judicial fiat. Thus, lacking an explicit statement of legislative intent to protect the specified plaintiff, there is no duty on the part of the public agency to perform inspection services. This interpretation of the duty obligation has served to protect public service agencies from liability. However, it has been the subject of expanding judicial criticism and has been rejected in some states. "The doctrine also has the effect of removing a great deal of the incentive for public bodies to see that the functions of government are carried out responsibly and with reasonable care." (*Dufrene v. Guarino*)

In examining the statutory duty issue, the status of the plaintiff is a principal focus of inquiry. Examination is made to determine whether or not the plaintiff is a member of the class of individuals owed the statutory inspection duty. For example, where the statute requires annual inspections of nursing homes, nursing home patients and staff would be considered members of the class owed the duty of inspection and consequent protection; on the other hand, visitors and deliverymen probably would not be class members. Broader interpretation occurs where the statute requires inspections intended to protect individual members of the public, such as inspections in

---

[46] Sellers, *op. cit.*, p. 103

The following states are reported as following the majority rule: Alabama, Arizona, Illinois, Kentucky, Maryland, Massachusetts, Michigan, Minnesota, New Jersey, New York, Ohio, Texas, and Washington.

[47] "We hold . . . that a municipality does not owe any individual a duty of care merely by the fact that it enacts a general ordinance requiring fire code inspections or by the fact that it undertakes an inspection for fire code violations. A duty of care arises only when there are additional indicia that the municipality has undertaken the responsibility of not only protecting itself, but also of protecting a particular class from the risks associated with fire code violations."

areas of public assembly. "Where the duty is owed to the public generally, no action lies on behalf of an individual for failure [to perform an inspection], but where the duty is intended to benefit the individuals composing the public, failure to perform such duty gives rise to a cause of action in favor of anyone injured by such failure," states *Gordon v. Holt*. Thus, the fact that a duty is of a public nature, and benefits the general public, does not require the conclusion that a city or state cannot be found liable for negligent performance of an inspection. (*Stewart v. Schmeider*) This expansive view of the general duty obligation is held by only a small minority of states.

A contractual obligation also could serve to establish an inspection duty. For example, where a community or private entity contracts for another to provide inspection services, an inspection duty for the benefit of individual community members might exist. Depending upon the phrasing of the contractual obligation, the duty to inspect can be over and above any applicable code or statutory requirement. Also, despite an explicit statutory immunity provision covering negligent fire inspection, immunity protection traditionally is limited to actions arising in tort, but not in contract.[48] (*Williams v. City of North Las Vegas*)

A city's ownership of a property can create a duty to perform inspections with due care. As discussed in Chapter 6 and above, a city, as the owner or occupier of land, would have the duty to exercise due care as to the condition of its property for all who are on it lawfully. *Trianon Park v. City of Hialeah* says: "Once a governmental entity builds or takes control of property or an improvement, it has the same common law duty as a private person to properly maintain and operate the property." For example, in *Donnelly v. City of New York*, the plaintiff fire fighter was awarded $2.6 million for injuries sustained while searching the burning roof of a tenement for trapped victims. He fell through roof, landed on the sixth floor, and sustained multiple fractures and burn injuries. The city had taken over the building for nonpayment of taxes seven months before the fire. Multiple inspections of the building during that period failed to uncover the many structural defects in the roof.

Finally, a "special relationship" between the plaintiff and inspector can establish a duty. A "special relationship" is a judicial determination that a duty of reasonable care existed between the parties, based upon the specific facts of the case. Factors considered by the court include whether or not there has been direct, personal contact between the inspector and plaintiff; the nature and authority of the inspector and whether he possessed exclusive authority to require corrective action; the representations made by the in-

[48] Despite explicit state code provisions for statutory immunity for negligence in building inspections, the immunity was held applicable only in tort. Where a city ordinance requiring inspection embodies terms of a franchise contract, the duty flows from the terms of the contract, and the immunity clause is not applicable. (*Williams v. City of North Las Vegas*)

spector; the reasonableness of relying on those representations; whether any code provisions required the inspector to undertake further action; the nature and gravity of the harm that potentially could result from failure to mandate correction; and the proximity of the plaintiff to the hazard that ultimately resulted in the injury. As stated in *Ribeiro v. Town of Granby*, particular attention is directed to any circumstances where "the risk created by the negligence of [the] municipal employee [was] of immediate and foreseeable physical injury to persons who [could not] reasonably protect themselves from it."[49] "... If conduct has gone forward to such a stage that inaction would commonly result, not negatively in merely withholding a benefit, but positively or actively in working an injury, there exists a relation out of which arises a duty to go forward." (*Fitzgerald v. 667 Hotel*)

Example cases that have held the existence of a special relationship include:

1. *Campbell v. City of Bellevue*: An inspector called to a home observed a neighbor's hazardous installation of an underwater light in a creek. The electrical code required that the inspector immediately disconnect circuit wiring to prevent the light's further use. As entry could not be made to the neighbor's electrical control panel, the location was only "red tagged." The inspector indicated to the father of the family in the inspected home that code-required corrective measures would be undertaken. The inspector subsequently asked the caretaker of the neighbor's house to disconnect the electrical circuit but did not personally verify that this act was performed. Six months after the inspection, the neighbor's son fell into the creek and was severly shocked. The boy's mother was electrocuted during an attempt to rescue him. The city was held liable in damages to the decedents' family.
2. *Fitzgerald v. 667 Hotel*: Following the remodeling of a hotel, which required the piercing of weight-bearing walls for the installation of air conditioning equipment, numerous cracks in various walls were seen. There also were building movement sounds. When the city's building inspector was called, he noted the seriousness of the cracking and the potential danger but issued no citation, nor did he direct that the building be evacuated. He only directed the owners to contact their structural engineer to examine the problem. When the structural engineer

[49] A residential boarding house, required to have two means of egress from upper floors, was cited for having only a single exit. The citation never was enforced by the city despite the fire department's expression of severe danger from the arrangement. A subsequent fire resulted in the death of a young boy trapped on upper floors. The court held the city not liable where building occupants knew of the danger many months before the fire but did not avail themselves of the opportunity to rectify or avoid the danger.

did so, he noted the seriousness, and recommended four corrective alternatives to the owner and city. All four recommendations were disapproved by the city. However, continued occupancy of the building was permitted.

On a Friday, seven months after the first of many inspections, material movement was noted in the structure; particles of brick and dust were seen falling frequently; door frames were out of line; pressure was noted on sprinkler piping; and rumbling noises were heard throughout the building. A tenant advised the building manager to evacuate the structure, then called emergency services of the city building department where he was told to call back on Monday. The police and fire department were called but both refused to respond because nothing had happened. The building subsequently collapsed, resulting in the deaths of four people. Again, the city was held liable in damages.

To protect municipalities from inspection liability, various states have passed immunity statutes.[50] Although these statutes are ineffective in insulating communities from liability for contractually-provided inspection services, courts generally have sustained the immunities for tortious inspection liability. Statutory immunity can take many forms, from providing immunity only when an inspector fails to exercise due care, to immunity for all actions, whether negligent or intentional. (*Morris v. Marin*)[51] Immunity also may be limited to state inspectors, thus leaving municipal inspectors open to liability. Reference to a state's specific statute, and the judicial interpretations thereof, is required.

---

[50] Sections 75-6103 and 75-6104(j), Kansas Statutes Annotated, provide an exemplary statute: "A governmental entity or an employee acting within the scope of the employee's employment shall not be liable for damages resulting from ... failure to make an inspection, or making an inadequate inspection, of any property other than the property of the governmental entity to determine whether the property complies with or violates any law or regulation or contains a hazard to public health and safety."

Comparable inspection immunity statutes have been passed by: California—California Government Code Section 818.6 (West 1983); Indiana—Indiana Statutes Annotated 34-4-16.5 (11) (Burns 1978); New Jersey—New Jersey Statutes Annotated Section 59:2-1; Nevada—Nevada Revised Statutes Section 41.032 (1981); and Oklahoma—Oklahoma Statutes Annotated 51 Section 155 (13) (West Supp. 1983).

[51] "Significantly, unlike Section 818.51[permit immunity], section 818.6 [inspector's immunity] contains no language limiting immunity to situations in which an entity has discretion to determine whether or not a safety or health hazard is present. In light of this difference in language and the purpose of the section, the immunity afforded by section 818.6 would attach to both ministerial and discretionary conduct in this area."

In *Siple v. City of Topeka*, the Supreme Court of Kansas "considered the purpose of inspection laws in determining any applicable immunity. The court pointed out that inspection laws are enacted for the benefit of the public as a whole. They are regulations designed to safeguard the public against fraud and injury and to promote the public's welfare and safety. They also provide an opportunity for an authorized public officer to examine an item and determine whether the standards imposed by law are being met by the responsible property owner. Because the legislature felt such inspection activities should be encouraged, it removed any threat of tort liability."[52]

"Causation analysis"—the factual and proximity relationship between the breached duty and the injury—plays the second most important part in determining inspection liability. Typically, the courts will require the causal relationship to be direct, with limited interference of other, mitigating events to sustain recovery. Where, for example, an inspector actively misrepresents a condition, which immediately results in the plaintiff's injury, recovery will be sustained, as in *Smuullen v. City of New York*.[53] However, occurrence of any intermediate event caused either by the plaintiff or a third party, or simply the lapse of time, can serve to prevent recovery.

In summary, the following inquiries are advised to determine the scope and extent of a city's and an individual inspector's exposure to liability:

1. Determine all sources, and the extent of obligation, for provision of inspection services. Pay particular attention to all municipal and state codes, promulgated by building, fire, and other agencies, that could establish an inspection duty.
2. Determine whether or not there is a statutory regulation or municipal code duty to "follow through" on all code violations uncovered. Is this being done? How is this assured?
3. Determine whether or not the inspection department and its individual inspectors tell the public that they will follow up on all code violations uncovered, no matter how inconsequential. Do inspectors represent the building as being "safe" rather than "code complying"?
4. Determine the existence and extent of any statutory or common law immunity provided for inspection services. What is the extent and ap-

---

[52] Jones, Richard E. and Alisa M. Dotson, "Kansas Defines Inspection Within the Context of Tort Immunity," *Municipal Attorney*, Vol. 25, No. 5, Sept.-Oct. 1984, pp. 5-7

[53] A workman, relying upon representations of a city inspector that trench walls were stable and secure without code-required shoring, entered a trench and was crushed by its subsequent collapse. Although the city was under no common law or statutory duty to render inspections, the city was found liable based upon the theory of a special relationship between the inspector and the workman, and the close proximity of misrepresentation to injury.

plication of any such immunity? Does it cover failure to inspect as well as negligence in the execution of inspections, mandatory and elective?

Municipal and Inspector Liability for Negligent Issuance of Permit or Certificate of Occupancy

The issuance of permits is an integral element of a community's fire prevention activities. Ranging from permits for construction, fire alarm and suppression system installation, and hazardous operations to building occupancy, the permits are an essential element of the community's fire prevention regulatory program. Distinguished from inspection liability, can a community or fire inspector be held liable for negligence in the issuance of a permit? The jurisdictions are split, with the majority holding no liability. This is in keeping with their comparable reasoning for no liability for negligence in performing or failure to perform an inspection. A small, expanding minority holds contrary views—that the fire official must exercise due care in reviewing plans and issuing permits, for the benefit of the permit applicant and the general public.

A case exemplary of the majority view is *Hoffert v. Owatonna Inn Towne Motel*, which held that the city could not be held responsible for the death and injury of motel occupants in a fire. The incident occurred two weeks after remodeling. The casualties were due to defective design and construction of exit stairways. The court held that "building codes, the issuance of building permits, and building inspections are devices used by municipalities to make sure that construction within the corporate limits of the municipality meets the standards established. As such, they [the standards] are designed to protect the public and are not meant to be an insurance policy by which the municipality guarantees that each building is built in compliance with the building codes and zoning codes. The charge for building permits is to offset expenses incurrred by the city in promoting this public interest and is in no way an insurance premium which makes the city liable for each item of defective construction in the premises."

The minority view is that the city and inspector owe a duty of care in their review and issuance of permits and, lacking statutory or common law immunity, can be liable for negligence in the performance of such service. In *J&B Development v. King County*, the county and building official were held liable for negligent review of building site plans and issuance of a construction permit. This negligance necessitated complete resiting of a building unlawfully located too close to a right-of-way. The court held that "as a matter of common sense, a home builder should be able to rely on the [building official] to furnish accurate information as well as valid building permits .... The issuance of a building permit inherently implies that the issuing agent has verified that the proposed structure is in compliance with applicable code provisions. Moreover, in this case, the subsequent inspection and approval by the county inspector reinforced J&B's earlier reliance on the validity of the building permit. If a home builder cannot rely on an official building permit issued by the county, there is little reason for requiring one, unless it is viewed as merely another mechanism for gathering taxes."

Fraudulent activities of a building or fire inspector—specifically, misrepresentation of a building's compliance with building and fire codes—can be a basis for inspector liability for a building owner's or third party's injuries. The fraud may be through actual collusion or through nondisclosure of adverse facts, either constituting fraudulent misrepresentation. The inspector is under a duty to disclose such facts to those injured. Also, immunity statutes which would protect the inspector from liability for negligent performance of his duties typically will be found inapplicable. (*Cooper v. Jevne*)

A further cautionary note: there may be no duty of care regarding injured plaintiffs, due to the state's adherence to the majority rule holding the city and fire inspector's duty as only a public one. However, a duty of care may be owed to third parties, such as insurance companies which are obligated to pay for the defendant's losses. In *Garrett v. Holiday Inns*,—a case involving a fatal fire in a motel—the town was held proportionately liable to motel owners, developers, and lessee operators for negligence in issuing a building permit.[54] Where the town, with knowledge of blatant fire code violations, nevertheless approved changes in a motel's building plans and issued a certificate of occupancy representing the premises to be safe and free from defects, the town could be held liable for losses. Although the town's duty of care did not extend to the injured parties, it did extend to those who directly relied on the town's representations; the motel's compliance with all fire safety requirements was inferred when the town issued a building permit and certificate of occupancy. This finding is especially important where the owners and insurers have had no contact with the site and thus detrimentally relied upon the city's representations before investing in the property.

Abatement of Hazardous Condition

Distinguished from nonperformance or negligent performance of inspection services is the negligent order for abatement of a discovered hazard. Although there may be no liability for the above or for directing a hazard's abatement, issuance of a deficient abatement order can impose liability. Specifically, where the means or method to abate the hazard is prescribed by a municipal department, liability can attach. Once the department, under its discretionary authority, has elected to prescribe the means a hazard will be eliminated, the department must carry out its responsibilities in a prudent manner. This includes exercising due care with respect to those who will be immediately affected by the abatement actions.

---

[54] Without a special relationship, this jurisdiction would not hold the city liable for negligence in the performance of inspections. However, it held the community liable for negligent review of building plans and issuance of permits.

In *In Re M/T Alva Cape*, a ship loaded with naptha exploded in New York harbor following collision with another steamship. To recover the remaining naptha and save the ship, the shipping company requested permission from the Coast Guard and the city to perform salvage operations in a designated explosives anchorage within the harbor. After obtaining approval from the Coast Guard of the procedure to be used to recover the remaining naptha, a salvage company undertook the removal operations. The city's fire commissioner was dissatisfied with the procedures employed and considered them inadequate to assure public safety, so he directed an alternative procedure. Use of that procedure resulted in an explosion on the vessel and the death of four salvage workers.[55]

In holding the city liable for the loss, the court noted: "Once the Fire Commissioner elected to place *Alva Cape* and its agents under the compulsion of law, he owed them the duty of exercising reasonable care, taking into account the emergency nature of the circumstances in formulating the terms of his order. While a city is often under no duty to provide services and exercise due care initially, once it elects to act in a specific situation it must carry out its responsibilities in a prudent manner, exercising due care with respect to those who will be immediately affected by its actions. While it is true that the Fire Commissioner's order involved the exercise of his discretion and expertise, this alone does not render the City immune from liability if the order was negligently issued."

In summary, the measure of an inspector's conduct in the abatement of a hazard or nuisance should be one of "reasonableness." As an example, an ordinance declares it unlawful to permit a fuse in excess of 15-amperes in a lighting circuit; and another ordinance declares that the violation of any ordinance shall be deemed to create a nuisance which may be abated summarily. Nevertheless, no one would contend that a fire inspector is justified in removing a 20-ampere fuse from a lighting circuit powering only an electric refrigerator, with resultant damage to the food it contained. Rather, reasoned conduct would be merely to notify the occupant of the danger, to disconnect all other nonessential items from the electrical circuit, and to direct the occupant to replace the unlawful fuse with one of the proper size.

Inspection of Private Dwellings

Subject to issuance of appropriate notice, there is no doubt that a right to inspect a private dwelling could be considered a reasonable exercise of the

---

[55] It is important to distinguish this case from those involving rendition of inspection services. "This case does not concern a failure to provide services or to plan properly for the public generally. Rather, we have the issuance of a mandatory order directed to a named recipient and limited by its terms to a unique set of circumstances. The Fire Commissioner knew that the issuance of such an order would compel its recipient to follow its command or suffer legal sanction for disobedience."

police power. This is true where there is reason to believe that conditions exist which make the property a fire menace to the community. But where a residence is so isolated from other occupancies or properties that it would not be considered a hazard, even if on fire, the right of privacy of the dwelling's owners might be considered paramount to the police power. There must be a balance between the two competing interests—the police power for public safety, and the right of privacy for personal protection.

In *Mapp v. Ohio*, the United States Supreme Court imposed a strict constitutional ban upon unreasonable searches and seizures as a protection against an invasion of privacy. By implication, this ruling placed an equally strict quarantine upon the use in all courts, state and Federal, of any evidence secured through any constitutionally-prohibited intrusions on privacy. Such a ban is to "apply to all invasions on the part of government and its employees of the sanctity of a man's home and the privacies of life. It is not the breaking of his doors, and the rummaging in his drawers, that constitutes the essence of the offense; but it is the invasion of his indefeasible right of personal security, personal liberty and private property."

In *Frank v. Maryland*, the Supreme Court, after declaring valid an ordinance which authorized health inspections of dwellings, said: "Two protections emerge from the broad constitutional proscription of official invasion. The first of these is the right to be secure from intrusion into personal privacy, the right to shut the door on officials of the state unless their entry is

JAMES MADISON, fourth President of the United States, often is credited with being "the father of the Constitution." Based on his plan, the Constitution provides for the executive, legislative, and judicial branches of government, outlines their interrelationships, and serves as the fundamental law of the nation.

under proper authority of law. The second and intimately related protection is self protection: the right to resist unauthorized entry which has as its design the securing of information to fortify coercive power of the state against the individual, information which may be used to effect a further deprivation of life or liberty or property. Thus, evidence of criminal action may not, save in very limited and closely confined situations, be seized without a judicially issued search warrant."

In *Camara v. Municipal Court*, the United States Supreme Court further limited the authority of government to perform nonconsensual inspections. The court held that, except for inspections for which there was voluntary consent, or inspections of publicly accessible areas during reasonable hours of the day, no inspection of a residential occupancy is permissible unless authorized by a warrant. In extending this personal protection, the court reasoned that "even the most law-abiding citizen has a very tangible interest in limiting the circumstances under which the sanctity of his home may be broken by official authority, for the possibility of criminal entry under the guise of official sanction is a serious threat to personal and family security .... [Fire inspections] do in fact jeopardize self-protection interests of the property owner. Like most regulatory laws, fire, health, and housing codes are enforced by criminal processes .... Finally, as this case demonstrates, refusal to permit an inspection is itself a crime, punishable by fine or even jail sentence.

"Under the present system, when the inspector demands entry, the occupant has no way of knowing whether enforcement of the municipal code involved requires inspection of his premises, no way of knowing the lawful limits of the inspector's power to search, and no way of knowing whether the inspector himself is acting under proper authorization .... The practical effect of this system is to leave the occupant subject to the discretion of the official in the field. This is precisely the discretion to invade private property which we have consistently circumscribed by a requirement that a disinterested party warrant the need to search .... In summary, we hold that administrative searches of the kind at issue here are significant intrusions upon the interests protected by the Fourth Amendment, that such searches when authorized and conducted without a warrant procedure lack the traditional safeguards which the Fourth Amendment guarantees to the individual ...."

Note that nothing in *Camara* or subsequent decisions forecloses prompt inspections, without warrants, where the law traditionally has allowed emergency entry. Typical cases include seizure of tainted food, compulsory smallpox vaccination, health quarantines, destruction of tubercular cattle, and fires. Where the exigency is real and presents an immediate and direct threat of personal injury, the law traditionally has allowed warrantless entry to preserve and protect life and property.

Without the exigency, inspections can be undertaken only under authorization of a warrant. It is important to distinguish the "probable cause" showing required to obtain an inspection warrant from the "probably cause" showing required to obtain a criminal search warrant. As outlined in *City of Seattle v. See*, "In determining whether a particular inspection is reasonable—and thus determining whether there is probable cause to issue a warrant for that inspection—the need for the inspection must be weighed in terms of the reasonable goals of code enforcement.... Probable cause in the criminal law sense is not required. For purposes of an administrative search such as this, probable cause justifying the issuance of a warrant may be based not only on specific evidence of an existing violation but also on a showing that reasonable legislative or administrative standards for conducting an... inspection are satisfied with respect to a particular [establishment]."

Factors that a court will consider in issuing an inspection warrant include: an appraisal of conditions in the area, the nature of the danger to be eliminated, the nature of the occupancy and the potential for injury, the passage of time, and the code provisions being enforced. However, "reasonableness" will remain the ultimate factor.

In contrast, probable cause for a criminal search means more than a bare suspicion: it needs a showing of "substantial probability that certain items are the fruits, instrumentalities or evidence of a crime and that these items are presently to be found at a certain location."[56] Such warrants cannot be based upon an area-wide program to find evidence of fruits of crimes. Rather, a specific showing of the evidence's existence, presently, at a specified location, is the key distinction of the criminal search warrant.

Ordinances that form the basis of inspection programs cannot be phrased to make it necessary for a property owner to grant a property inspection to protect his property interests. The ordinance in a questioned case required that, should a property lie vacant more than six months, the building owner would be required to obtain a new certificate of occupancy for the structure, which could only be issued upon passage of a reinspection. All parties recognized that involuntary inspection of such properties could be performed only under authority of a warrant. However, the city argued at length that the ordinance involved only consensual entries and that reinspections would be performed only with the consent of the owner. The court held the city's argument to be specious, finding: "To compel a property owner to let his property lie vacant and to prohibit him from selling it, unless he 'consents' to a warrantless search, is to require an involuntary consent. The owner's basic right to use and enjoy the fruits of his property cannot be conditioned on his waiving his constitutional rights under the Fourth Amendment...." (*Currier v. City of Pasadena*)

---

[56] Kamisar, Yale et al., *Modern Criminal Procedure*, West Pub. Co., 5th Edition, 1980, p. 268

Inspection of Commercial Premises

Inspection of commercial premises involves ramifications similar to those associated with private dwellings. A case in point, *See v. City of Seattle* involved a warehouse which the fire department wanted to inspect. The inspection authority was a local ordinance granting the fire chief the right "to enter all buildings and premises except the interior of dwellings as often as may be necessary" to determine compliance with fire code requirements. The owner refused to permit an inspection of his commercial premises on the grounds that the ordinance was invalid. He said the fire chief had no search warrant, nor any probable cause to believe that a violation of any law existed on his premises. In reversing the owner's conviction for failure to permit an inspection, the United States Supreme Court followed its reasoning in *Camara* and held: "The businessman, like the occupant of a residence, has the constitutional right to go about his business free from unreasonable entries upon his private commercial property."

"We therefore conclude," the Supreme Court wrote, "that administrative entry, without consent, upon the portions of commercial premises which are not open to the public may only be compelled through prosecution or physical force within the framework of a warrant procedure. We do not in any way imply that business premises may not reasonably be inspected in many more situations than private homes, nor do we question such accepted regulatory techniques as licensing programs which require inspections prior to operating a business or marketing a product.... We hold only that the basic component of a reasonable search under the Fourth Amendment—that it not be enforced without a suitable warrant procedure—is applicable in this context, as in others, to business as well as to residential premises. Therefore, appellant may not be prosecuted for exercising his constitutional right to insist that the fire inspector obtain a warrant authorizing entry upon appellant's locked warehouse."

Inspection Program and Procedures

As a result of the above decisions, it is advisable to establish a definite program for conducting inspections and to set forth in a procedures manual the basis and frequency for making inspections for purposes including prefire planning, code enforcement, and permit issuance. With the exception of such special occupancies as schools, hospitals, sanitariums, and nursing homes, which require the owner to permit entry in order to maintain his permit, all routine inspections should be programmed on the basis of an orderly geographical approach. Inspections can be planned so that fire fighters will proceed through their districts on a building-by-building, block-by-block basis. Then when that rare objection is voiced by a building owner who does not want his premises inspected, it can be shown that his building is not being singled out for a shakedown, but was reached as a part of the planned sequence of inspections.

The importance of this system is aptly demonstrated by the case of *City of Seattle v. See*. Here, a fire inspector visited the defendant's apartment complex, noted a deficiency in fire extinguishers, and cited the owner. Numerous attempts to contact the owner followed, in person and by telephone, in an effort to obtain the owner's permission to perform a follow-up inspection to verify correction of the code deficiency. The inspector was told to "look [defendant] up in the United States Supreme Court and Fire Department records," and was refused permissive entry to perform the reinspection. A warrant was obtained and a reinspection conducted, resulting in another citation, prosecution of the owner for ordinance violation, and his conviction. In sustaining the conviction, the court held: "Reasonable legislative and administrative standards for conducting an inspection at See's building were followed. The Fire Department Inspection Manual outlines specific procedures which the Fire Department must follow. It provides for annual inspections of all occupancies, except private dwellings. The defendant's building was part of an overall inspection program for Seattle."

Right of Entry Ordinances

Following the *Camara* and *See* decisions, a number of cities amended their ordinances relating to a fire inspector's right of entry. It is interesting to note these changes made in response to concerns expressed by the Supreme Court Justices. For example, one major city amended its ordinance to read as follows:

1. Any fire inspector, when engaged in fire prevention and inspection work, is authorized at any and all reasonable times to enter and examine any building, vessel, vehicle, or place for the purpose of making inspections. Except when an emergency exists, before entering a private building or apartment the inspector shall obtain the consent of an occupant thereof or a warrant of the Municipal Court authorizing his entry for the purpose of inspection. As used in this section, "emergency" means unforeseen circumstances which the inspector knows or has reason to believe exist, and which reasonably may constitute an immediate danger to public health or safety.
2. The official badge and uniform of the Bureau of Fire, when lawfully worn, shall authorize an inspector of the Bureau of Fire to enter and inspect buildings, vessels, vehicles, and premises as herein set forth.
3. It is unlawful for any person to interfere with or attempt to prevent any inspector from entering and inspecting any building, vessel, vehicle, or premises, or, when an emergency exists or the inspector exhibits a warrant authorizing entry, from entering and inspecting any private building or apartment.

4. It is unlawful for any unauthorized person to wear the official badge or uniform of the Bureau of Fire, or to impersonate a fire inspector with the intent of unlawfully gaining access to any building, vessel, vehicle, or premises in the city.

Administrative Inspection Warrants

In addition to revising local entry statutes, the impact of *See* and *Camara* has been the adoption in many states of statutes that prescribe the procedures to be followed and the allegations necessary to obtain an inspection warrant. In California, the procedure for issuing an inspection warrant is contained in Sections 1822.50 through 1822.57 of the Code of Civil Procedure. Those sections provide for a judge of a court of record to issue a warrant of inspection, on the basis of an application, in affidavit form, showing "cause" for the desired inspection. In Section 1822.52 the California Legislature has defined the "cause" which must be shown:

"Cause shall be deemed to exist if either reasonable legislative or administrative standards for conducting a routine or area inspection are satisfied with respect to the particular place, dwelling, structure, premises, or vehicle, or there is reason to believe that a condition of nonconformity exists with respect to the particular place, dwelling, structure, premises, or vehicle."

In addition, the times of day that inspection warrants can be executed and under what procedures are strictly limited. Specifically, before carrying out a routine inspection under provisions of a warrant, permission for inspection must first be requested and refused; then 24 hours' notice must be given to the owner/occupant before the warrant can be executed. Inspection by force is prohibited. As stated by the *Camara* court: "In the case of most routine area inspections, there is no compelling urgency to inspect at a particular time or on a particular day. Moreover, most citizens allow inspections of their property without a warrant. Thus, as a practical matter and in light of the Fourth Amendment's requirement that a warrant specify the property to be searched, it seems likely that warrants should normally be sought only after entry is refused unless there has been a citizen complaint or there is other satisfactory reason for securing immediate entry. Similarly, the requirement of a warrant procedure does not suggest any change in what seems to be the prevailing local policy, in most situations, of authorizing entry, but not entry by force, to inspect."

Except as noted in *King County v. Primeau*, courts can hold that no advance notice is required under certain conditions: where the warrant is sought in response to the existence of a known hazard or dangerous condition; upon a citizen's complaint; when the initial inspection was refused; when code deficiencies were uncovered; when permission for reinspection is not reasonably anticipated; or when there is other satisfactory reason for securing immediate entry.

In the appendix to this chapter are sample copies of forms for an inspection warrant (Document 1), affidavit (Document 2), and notice to occupant (Document 3). Reference to these documents is suggested for a better comprehension of what facts must be specifically alleged in order to sustain a magistrate's issuance of an inspection warrant.

Search and Seizure

Frequently, it is the duty of fire inspectors to confiscate evidence of a crime or to apprehend violators of the law. Because of this, it is well for them to keep in mind the limitations against unreasonable searches and seizures provided for by the United States Constitution and the constitutions of the various states.[57]

The fundamental rule is that, subject to very limited exceptions, any search or seizure must be done under authority of a warrant, based upon "probable cause" and issued by a neutral magistrate. Where a search or seizure is performed without warrant, and does not fall within any exception to the warrant rule, it is considered constitutionally "unreasonable," with any resulting evidence prohibited from consideration in a court of law. The essence of the test of reasonableness is the balancing of fundamental interests—those of the person (to be free from unreasonable searches and seizures), and those of the governments (to seize unlawful materials, to obtain evidence of illegal acts, and to uphold the laws). In each case, has the government presented enough reasoned facts to justify limiting a person's constitutional protections and permit search and seizure? Should a person's reasonable expectation of privacy and freedom from unreasonable searches and seizures be limited?

It is important to note that an individual's right to privacy, including prohibition of unreasonable search or seizure, is not a property-oriented right. It is not a right limited within geographic boundaries. Rather, the Fourth Amendment's protections are personal; they protect people, not places. "What a person knowingly exposes to the public, even in his own home or office, is not a subject of Fourth Amendment protection," holds *Katz v. U.S.* "But what he seeks to preserve as private, even in an area accessible to the public, may be constitutionally protected." *Jones v. U.S.* says, "A person, of course, can have a legally sufficient interest in a place other than his own home so that the Fourth Amendment protects him from unreasonable governmental intrusion into that place."

Where any search or seizure is "unreasonable," courts hold that government should not be permitted to benefit from such unreasonable action. This

---

[57] "The right of the people to be secure in their persons, houses, papers, and effects, against unreasonable searches and seizures, shall not be violated, and no warrants shall issue, but upon probable cause, supported by oath or affirmation, and particularly describing the place to be searched, and the person or things to be seized." Amendment IV (1791), United States Constitution.

is the premise of the "exclusionary rule," whose purpose "is not to redress the injury to the privacy of the search victim.... Instead, the rule's prime purpose is to deter future unlawful police conduct and thereby effectuate the guarantee of the Fourth Amendment against unreasonable searches and seizures.... In sum, the rule is a judicially created remedy designed to safeguard Fourth Amendment rights generally through its deterrent effect, rather than a personal constitutional right of the party aggrieved." (*U.S. v. Calandra*)

Court decisions addressing issues of "probable cause," the scope and extent of searches, and the "reasonableness" of warrantless searches and seizures under perceived "exigent" circumstances are myriad. The rules resulting from these decisions are varied, based upon the amalgam of protections provided by the Federal and state constitutions. However, some of the basic exceptions permitting warrantless searches and seizures follow.

Where an inspector has made a lawful arrest, with or without an arrest warrant, he has an incidental right of search and seizure—to protect the accused and the arresting officer, and to confiscate evidence. Note that such search and seizure is limited to evidence on the person and within the immediate "area of control" of the suspect: this might be an area consisting of the arrestee's person and the area that can be reached to obtain a weapon or to destroy evidence. (*Chimel v. California*) A search of the house of the arrested party would not be permissible without a search warrant.

A second exception to the warrant requirement is that evidence of a crime may be seized when it is in "plain view" of the officer. As long as the officer is legally at the viewing place, he is empowered to seize such evidence. (*Coolidge v. New Hampshire*) The seizure of evidence of arson, following extinguishment of a fire, is a typical example. "A burning building presents a clear exigency that renders a warrantless entry reasonable and, once in the building, fire fighters may seize evidence of arson that is in plain view and they may investigate the cause of the fire," under *Michigan v. Tyler*.

South Dakota v. Jorgensen says this is true where a fire cause investigation immediately follows fire extinguishment; or, in the case of a nighttime fire, after the latent smoke clears and daylight arrives.

Under certain circumstances, courts hold that the exigency that permitted the initial warrantless entry to extinguish the fire continues through to the start of fire cause investigation. Rejecting the proposition that the exigency that justifies warrantless entry ends with the extinguishing of the flames, the [United States Supreme] Court went on to hold: "We think this view of the fire fighting function is unrealistically narrow, however. Fire officials are charged not only with extinguishing fires, but with finding their causes. Prompt determination of the fire's origin may be necessary to prevent its recurrence, as through the detection of continuing dangers such as faulty wiring or a defective furnace. Immediate investigation may also be necessary to preserve evidence from intentional or accidental destruction. And, of course, the sooner the officials complete their duties, the less will be their subsequent interference with the privacy and recovery efforts of the victims." For these reasons, officials need no warrant to remain in a building for a reasonable time to investigate the cause of a blaze after it has been extinguished. And if the warrantless entry to put out the fire and determine its cause is constitutional, the warrantless seizure of evidence while inspecting the premises for these purposes is constitutional.[58]

Cases

The owner of a home that was set on fire was charged with arson. The trial court suppressed all evidence obtained by the fire investigator except for photographs of the building's exterior. The fire was extinguished between 9:00 and 9:30 p.m.; however, due to residual heat, dripping water, and darkness, initial investigation was postponed until first light. The next day, the investigator arived at 9:00 a.m., took floor samples and interior photographs, and concluded that the fire was of incendiary origin. The appellate court reversed this conclusion, holding that the early-morning investigation was but a continuation of the prior evening's original entry; that the exigency justifying the initial warrantless entry remained; and that the interior pictures and floor samples were admissible at trial. (*Commonwealth v. Smith*)

In *State v. Zindros*, the owner of a pizza restaurant that was the site of a fire of suspicious origin was convicted of arson. The arson finding was based upon evidence obtained during multiple entries of the restaurant, ranging in time from immediately after the fire's extinguishment to 11 days after the fire. Although the initial investigations following the fire were to determine its cause and origin, the subsequent entries were to gather evidence; the fire was determined to be of incendiary origin, with the pizza parlor owner the prime suspect. The court suppressed all evidence of the subsequent investigations. Once the premises were left after conclusion of the initial investigation, said the court, the owner had a reasonable expectation of privacy in his damaged facility and any subsequent entry could be undertaken only under authority of a search warrant.

Right to Photograph

There is no question that fire investigators may take photographs during and following extinguishment operations. These pictures serve as incidental

---

[58] The defendant's conviction for arson was upheld where a deputy fire marshal entered the scene of a fire approximately two hours after its extinguishment. Evidence seized during that investigative entry was held admissible into evidence. However, evidence seized during subsequent investigative site visits was suppressed, as the "exigency" to allow initial entry—the blaze—had long since passed.

See also: *Michigan v. Clifford* regarding continuation of exigency to permit fire cause investigation upon the clearance of smoke conditions and arrival of first light.

tools in establishing the cause and origin of the fire, or can be used in evidence to establish the commission of a crime. However, where a fire has not yet occurred, and the investigator's only purpose for entering the premises is to photograph fire hazards, valid objections may be raised by the occupant. In the absence of specific provisions in the ordinance or statute granting fire inspectors the right to enter, inspect, and photograph premises for fire hazards, it is highly doubtful that they have the right to take pictures. Note also that rights of property owners for the protection of proprietary information may override the fire department's authority to take pictures, even of fire hazards. Particular caution is advised.

Fire Inspection Records

Disclosure of the records and reports of fire inspections and fire investigations generally is governed by state public disclosure laws. For example, Kansas Statutes Annotated 45-201(a) provides: "All official public records of the state, counties, municipalities, townships, ... districts, ... which by law are required to be kept and maintained, except ... records specifically closed by law or by directive authorized by law, shall at all times be open for a personal inspection by any citizen, and those in charge of such records shall not refuse this privilege to any citizen."

As most model fire codes contain a provision requiring promulgation and retention of inspection and investigation records, the adopting fire department thus can come under an obligation to publicly disclose such records.[59] Thus in *City of Topeka v. Stauffer Communications*, where the fire department sought not to disclose inspection records of high-rise motel buildings for use by the local newspaper, the court directed the records' release. The court held that unless the department could prove that the materials were privileged, such as between the city and its attorneys or as part of a continuing criminal investigation, the materials must be released under public disclosure laws. Such disclosure must include not only the formal record, but also interdepartmental memoranda and reports, correspondence with owners, and communications with the public.

The disclosure obligation is not limited to full-time professional departments; it extends to any volunteer department that receives any financial support, directly or indirectly, from public sources. Thus, in *Schwartzmann*

---

[59] For example, see Uniform Fire Code, 1985 edition, Section 2.203: "(a) The fire department shall keep a record of all fires occurring within its jurisdiction and of all facts concerning the same, including statistics as to the extent of such fires and the damage caused thereby, together with such other information as may be required by the chief. (b) The fire prevention bureau shall retain for not less than three years a record of such inspection and investigation made showing the cause, the findings and disposition of each such inspection or investigation."

v. Merritt Island Volunteer Fire Department the volunteer fire department—which used county apparatus and received funding for protective equipment—was obligated to disclose its membership files, minutes of its meetings, and records of its charitable fund-raising activities.

The policy reasoning was outlined in *Westchester Rockland Newspapers v. Kimball* by Justice Jacob D. Fuchsberg of the New York Court of Appeals: "Successful implementation of the policies motivating the enactment of the Freedom of Information Law centers on goals as broad as the achievement of a more informed electorate and a more responsible and responsive officialdom. By their very nature such objectives cannot hope to be attained unless the measures taken to bring them about permeate the body politic to a point where they become the rule rather than the exception." Here the city was required to disclose records of a public lottery sponsored by the volunteer fire department.

The burden of demonstrating that an exemption from disclosure applies falls squarely upon the shoulders of the one asserting the privilege.

Review Activities

1. Write a brief comparison between negligence and strict liability. Illustrate your description by citing examples of each.
2. Describe a "special relationship." How could it be utilized to create a special duty? From your own experience, describe an instance where a special duty could have been created. What could you have done to preclude such a finding?
3. What is a "nuisance"? What factors would you consider in making such a finding? How would you abate it?
4. What are "police powers"? Who possesses them? What is the scope and limitation of their reach?
5. Describe the means in your community for adoption of a new fire code. What agencies are involved and what activities do they perform in the process?
6. A. What rights does a fire inspector have in inspecting private dwellings? Commercial buildings?
   B. What is the proper and legal way for a fire inspector to conduct an inspection?
7. Distinguish among Federal, state, and local regulation in the field of fire prevention. Cite an example of jurisdictional conflict other than the one mentioned in the text.

Appendix to Chapter Eight

Document 1

SUPERIOR COURT OF THE STATE OF (state name)
FOR THE COUNTY OF (county name)

In the Matter of the Application
of (fire department name)

No. (case number)
INSPECTION WARRANT

The People of the State of (state name) to: (police officer/fire inspector, etc.)

Upon good cause shown to the Court, you are hereby commanded to conduct an inspection, as (authorized) (required) by (code section), of the premises (or vehicle) described as (premises/vehicle description) for the purpose of (reason for inspection).

This warrant is effective from the date hereof for a period not to exceed 14 days and it shall be returned to the judge whose signature is affixed below.

Said inspection shall not be made between 6:00 P.M. and 8:00 A.M. of the succeeding day (nor in the absence of the owner or occupant). Notice shall be given to (owner) (occupant) at least 24 hours before this warrant is executed.

Dated: _____, 19_____, at _______ __.M.

(signed)
Judge of the Superior Court

Document 2

SUPERIOR COURT OF THE STATE OF (state name)
FOR THE COUNTY OF (county name)

In the Matter of the Application of (fire department name)

No. (case number)
AFFIDAVIT FOR INSPECTION (Administrative Search) WARRANT

(Name of inspector/officer), being duly sworn, deposes and says:

1. That he is the (describe official capacity).
2. That pursuant to (code section) he is required (or authorized) to conduct an inspection of the (place, dwelling, structure, premises or vehicle) described as (place/vehicle description).
3. That the purpose for which said inspection is sought is (identify purpose).
4. That (name of owner/occupant) of (number, street), (city), (state), is the owner (or occupant) of said (premises/vehicle).
5. That on the _____ day of ______, 19____, affiant requested permission of said owner (or occupant) to conduct an inspection as required (or authorized) by the law or regulation aforesaid and consent thereto was denied.

(or)

(Explain why consent not sought).

Wherefore, affiant prays that an order of Court be made issuing a warrant for inspection of the (premises/vehicle) to be made by him.

Subscribed and sworn to before me, this _____ day of _______, 19_____.

(signed)
____
Notary Public in and for said County and State

Document 3

Mr. (owner/occupant)
(street)
____
(city) ________, (state) (ZIP code)

24-HOUR NOTICE
INSPECTION WARRANT

Notice is hereby given to you as owner (occupant) of the premises (or describe) at (premises address/vehicle description), that on the _____ day of ________, 19___, Judge (name) of the Superior Court of the State of (state), County of (county), issued a warrant for the inspection of the above described (premises/vehicle) based upon your refusal to allow such inspection.

Inspection is sought for the purpose of (verifying compliance, etc.) pursuant to (code section).

You are further notified that unless such inspection is permitted within 24 hours of the issuance of said warrant, it will be executed by the undersigned under order of court. Your refusal to permit an inspection as authorized by the warrant is punishable by law, under Section __________ of the Code of Civil Procedure of the State of __________, as a misdemeanor.

Dated: ______________

(signed) ______________________________

(title) ______________________________

Chapter Nine

Pensions

"Pension" and "Compensation" Distinguished

Despite recurring references to fire fighters' retirement pensions as "delayed compensation," the words "pension" and "compensation" are not synonymous. A pension is but one element of the compensation package (salary, vacation, disability, and retirement benefits) provided to public employees. It is but one of the items of compensation considered by the prospective public employee in deciding whether or not to take a job. A pension also is distinct from workers' compensation, which provides substitute salary or wages in the event of work-related injury. Thus, an award for injuries under a workers' compensation act will not prevent the recipient from becoming a beneficiary under a pension plan. However, the combined amount of monies received under a combined pension and workers' compensation award cannot exceed statutory limits.

For example, in a 1965 California case involving a fire fighter, the City of Los Angeles claimed that it did not have to pay workers' compensation claims as long as it paid pension benefits. The local union contested this stand and won. The California Supreme Court held that the city would be liable for workers' compensation claims plus that portion of the pension which had been contributed by the individual involved. As the fire fighter had paid 6 percent of his salary into the pension fund and the city paid an amount equal to 13 percent of his salary, the fire fighter was entitled to his proportional contribution to the plan.

Pension Systems

In general, three types of pension plans are used by local government. First, where provided by state statute or authorized by charter, a city is empowered to organize its own pension program or participate in a multi-jurisdiction state program. In addition, a state can mandate the provision of a pension plan by a local government employer. Second, cities may participate in the Federal Social Security retirement program. As the trend has been to increase participation in the Federal program, and in light of the recent United States Supreme Court decision holding that the Federal government can regulate conditions of public employment, greater participation by local government in the Federal Social Security System is envisioned. (See *Garcia v. San Antonio.*) Finally, private benefit and relief associations can independently manage selected pension programs for municipal employees.

Constitutionality of Pension Plans

Retirement and disability pensions for fire fighters have been sustained despite various constitutional attacks. Because pensions are considered one element of the compensation package offered to fire fighters, retirement benefits have been attacked as unlawful gifts of public funds or unlawful increases in public compensation. As stated in one court challenge, *San Francisco Fire Fighters v. City and County of San Francisco*: "It is well settled that a public employee's pension constitutes an element of compensation, and a vested contractual right to pension benefits accrues upon acceptance of employment."

What happens when a public employee, originally hired by one city department and covered by that department's separate pension program, transfers to another department within the city? Specifically, what happens where the city employs a by-department, rather than a city-wide, pension plan system (such as a Police Department/Fire Department Pension Plan versus an All City Employees' Pension Plan)? What happens when, in the interim between initial public hire and transfer, the city's pension program is changed, reducing an employee's future pension benefits? Is the transferred individual entitled to the higher pension that would have accrued had he been hired by his current department? Careful analysis of the vesting provisions and transfer requirements of the respective pension plans is required. The case of *San Francisco Fire Fighters v. City and County of San Francisco* was brought by former police officers who transferred to the fire department. They challenged pension program provisions that said transferred employees were subject to the same provisions as newly-hired city employees—not provisions in force at the time of initial public employment. Where calculation of retirement pension benefits was based upon the dates of hire by the specific city department, and not dates of initial city employment, former officers were entitled only to benefits calculated under the newly-hired plan. In addition, the officers were not entitled to a current pension, and the police department's plan had explicit requirements for transfer of accumulated time and contributions to other departments' programs. As a result, the officers became subject to provisions of their new department's plan in effect at the time they transferred.

Pension Provisions Liberally Construed

Since pensions are a matter of statute, the person who seeks a pension must show that he meets all the prerequisites to be entitled to the benefit. These include such matters as minimum years of service, disability in the line of duty, proper and timely presentation of his claim, and compliance with pension fund contribution requirements. Pension boards have a fiduciary duty to assure strict compliance with all terms governing application of the pension program. "One line of authority holds that statutes providing for pensions ought to be construed liberally in favor of the persons intended to be benefitted thereby, but the better rule is to construe such statutes in light of the general plan evidenced by the enactment including protection of the solvency of the fund and protection of the interests of the municipality and its taxpayers."[1]

As a general rule, pension statutes are to be liberally construed in favor of the pensioner. Some courts, however, are empowered to consider the consequences of an interpretation of a pension rule or statute whenever more than one interpretation is possible. Such courts are bound to consider the "public interest as against any private interest" in the interpretation of pension regulations. For example, in *Borough of Beaver v. Liston*, a pension plan was held not to take effect retroactively. Thus it did not include fire fighters who had left the fire department at the time of the plan's enactment. The court ruled that no retroactive application was provided by statute, and that such an interpretation would be contrary to the public interest.

Personnel Entitled to Pension

A recurring question regarding fire fighter pension plan eligibility is whether or not an individual is a "member" of the fire fighting force. Within the meaning of a disability or retirement pension law, is he entitled to a "fire fighter's" pension? Through liberal construction of fire department pension provisions, the following have been held to be entitled to the plans' benefits: blacksmith helper, extra man, janitress, veterinarian, lineman, watchman, fire alarm crew, fire hydrant maintenance personnel, fire department secretary, and clerk.

It may seem from this list that actual fire fighting experience is not the criterion for determining whether or not a fire department employee is entitled to share in the fire fighter's pension program. This position is not universally held, however. In sustaining the denial of such benefits to San Francisco fire boat pilots and engineers, a California appeals court held in *Carrich v. Sherman*: "Relief and pension systems for firemen are to be liberally construed for the benefit of firemen, but it is of no advantage, and is in fact a

---

[1] Sands, C. Dallas and Michael Libonati, *Local Government Law*, Callaghan and Co., 1985, Section 10.34, p. 10-224

disadvantage, to have included within the pension system a number of men who are not exposed to danger." Likewise, in the Massachusetts case of *Elliott v. City of Boston*, members of the repair division of the Boston Fire Department were held ineligible for a retirement pension on the ground that "members" of the fire department, within the meaning of the statute, include only those fire fighters who wear a uniform and have attended drill school. The court noted that the essential difference in the work of the repair men and the fire fighters lay in the fact that the fire fighter's work is considered hazardous. The New Jersey case of *Taggart v. City of Asbury Park* concerned men in the public safety department who, after a governmental reorganization, continued to perform the same fire fighting duties as when they were members of the fire department. Under these circumstances, the court ruled that the men were entitled to fire fighter's pensions.

The Supreme Court of Pennsylvania held that a fire inspector was a "regularly appointed member" of the City of Washington, Pennsylvania, Fire Department, and hence was entitled to a pension. In its finding, the court rejected a limited interpretation put on the definition of a "fireman" by the Fireman's Relief Association. This definition described a fire fighter as a person who is actually engaged in fighting and extinguishing fires. In holding this to be too limiting, the court wrote: "How this definition leads the Association to conclude that 'fighting fires' is limited to fighting them by pulling hose, squirting water or driving a truck is hard to fathom. Fire fighting has become more and more complicated in the modern technological world."[2]

The test generally applied in determining pension eligibility is whether or not the activities of the individual have a close relationship to the mission of the fire department. Thus, the determining factor is not the origin of the pension fund nor work combating flames, but the nature of the employee's duty. That is, does he directly—not remotely—carry out a function assigned to the fire department by ordinance or charter provision? It is not reasonable to contend that in a large municipal fire department, directed by law not only to extinguish fires but also to contain nonfire hazardous conditions, only those who come in contact with the smoke and flames are "members" of the fire department. It does not require a liberal interpretation to concede that arson investigators, rescue squad members, salvage crews, hazardous materials specialists, and fire apparatus operators are all engaged in regular activities of the fire department and should be considered department "members." However, it would be beyond the bounds of liberal interpretation to include as "members" the secretarial force, apparatus maintenance people, or facility repair personnel who are only remotely assisting in carrying out the fire department's mission.

---

[2] Morse, H. Newcomb, "Legal Insight: Pension Rights," *Fire Command*, Vol. 41, No. 7, July 1974, p. 32

Determination of Pension Amount

In general, the amount of the pension benefit to which an individual is entitled is outlined by charter, statute, or rules of the pension board. Careful analysis of the statute or regulations governing plan administration is required. As a rule, however, pension plan payments are based upon the base salary of the individual, without consideration of any incentive pay or allowances. In addition, "Appellate courts of [most] jurisdictions construing the meaning of the term salary as employed in . . . pension laws have uniformly held that overtime payments are not salary for purposes of computing monthly pension amounts." (*Borough of Beaver v. Liston*) Thus, selection of items for inclusion in or exclusion from pension benefit calculation requires critical examination of the applicable regulations.

For example, in *Hill v. City of Lincoln*, a class action lawsuit was brought by active city police and fire fighters, who challenged withholding amounts for the pension plan. Benefits were to be based upon amounts received as "salary" and contributions based upon an employee's "regular pay," but there was confusion as to what payments were to be considered for each category. Plaintiff employees contended that the statutory term "salary" meant only "basic pay," whereas the term "regular pay" included "basic pay" plus all extra compensation, such as for overtime, holiday service, and college credits. "In other words [plaintiffs insisted], deductions are to be based on the lesser pay, and benefits on the greater." In rejecting this contention, the court examined the actuarial balance between plan income and payments, as well as the prior practice of the community. The court made these studies to assure the financial basis and stability of the plan, saying "It is axiomatic that there is no such thing as 'free money.' "

What happens where an employee, considering retirement, asks the benefits officer if he would be entitled to a pension of at least X amount, based upon his work history? The stated amount was considered the minimum necessary for the employee to safely retire and still cover all anticipated obligations. Following his retirement, he is informed that the benefits officer miscalculated, and only a lesser amount can be authorized as retirement pay. Can the retiree force payment of the higher amount? Can the city recover for the erroneously provided higher payments? In *Kern v. City of Flint*, this series of events actually occurred. The court held: "We agree with defendant's contention that generally the rule in this jurisdiction is that where one pays money to another by mistake, that person is entitled to recover the amount of the overpayment even if the mistake is due to lack of investigation. However, an exception to the rule exists where the party receiving the money has changed his position in reliance upon his receipt of it so that it would be inequitable to permit recovery." Thus, because Mr. Kern had materially changed his position, by retiring and thereby reducing his ability to maintain his income level, the city could not seek return of the excess pension payments or pay anything lower than the originally stated amount.

Death or Injury in Line of Duty

In order to establish a right to a disability pension, or to benefits granted widows and dependents of fire fighters killed in action, it is necessary for the claimant to prove that the injury or death occurred in line of duty or as the result of an incident happening within the scope of the fire fighter's employment. However, the question of what activities should be considered "in the performance of duty" or "work related"[3] is a matter of judicial controversy. Settlement of the question in any given instance can only be by examination of local pension board policy and judicial decisions interpreting pension provisions.

Disability Determinations

The responsibility for determining who is entitled to a disability pension typically is vested in a pension board, which may or may not be an entity of local government. This same board also is responsible for regulating service pensions. Regular service pension cases usually are not items of major challenge to the board, nor are such cases subject to significant judicial commentary. Therefore, review of pension board activities in this area is not included here. On the other hand, disability pensions—service- and nonservice-related—typically are the subject of greatest fire fighter concern, so it is appropriate to review the scope of pension board responsibility, particularly in this area.

Pension board duties usually are prescribed by statute, city ordinance, or charter. Typical items for determination include: "1. whether a [pension] system member has become physically or mentally incapacitated; 2. whether such physical or mental incapacitation results from injuries received or sickness caused by the discharge of the member's duties; and 3. whether the member is incapable as a result of the . . . incapacity of performing his duties."[4]

By what standards does a board determine if a disability pension is warranted? "In the determination of whether the board owes a duty to grant a disability pension, the board must decide whether a [pension] system member has become physically or mentally incapacitated. The degree of incapacity warranting the grant of a disability pension is defined, however, by the third factor—whether the physical or mental incapacity results in the system member's inability to perform his duties."[5]

A disability pension cannot be denied on unreasonable or arbitrary grounds. For example, a San Antonio fire fighter became unable to perform

---

[3] A "work related injury" can arise from employment that brings the employee into a position of danger from injurious forces even though the employer had no connection with or control over the forces and could not have foreseen the risk.

[4] Charter of the City of Los Angeles, Section 190.12

[5] *Ibid.*

his duties because of an injury which made his little finger stiff. He underwent an operation to correct the condition, but the surgery was unsuccessful. A medical panel then found that, unless his finger was amputated, he would be totally and permanently disabled and could not carry on his duties. When the fire fighter refused to have the amputation, the pension board denied him a disability pension. The court of civil appeals overruled the board's decision, saying that the pension board had no authority to order such an operation and could not penalize the fire fighter for refusing to have it.[6]

When a pension board's denial of a disability pension goes against the overwhelming evidence in support of the pension, the ruling will be overturned by the courts. The case history of a Houston fire fighter highlights such an event. His entrance medical examination showed that he was in good physical condition, and another examination when he returned to the fire department after military service also showed him to be in good physical shape. However, after he was thrown 30 feet from the fire truck when it collided with an ambulance, his private physician found that the injuries to his back had resulted in his permanent disability. However, the city physician said that the man was able to return to duty. The fire fighter resumed work, then, while loading wet hose onto a fire truck, again injured his back. This caused him to fall, making his condition worse. Medical examination disclosed that his back was severely injured, with accompanying atrophy of leg muscles. Even with a back brace, the fire fighter remained in severe pain. Because the city physician again examined him and said he was fit for duty, the pension board denied his application for a disability pension. In affirming the district court's ruling that the fire fighter was entitled to a pension, the appellate court found that the evidence did not reasonably support the retirement board's decision in light of the case's record as a whole. A writ of mandate by the court directed the board to issue a disability pension.[7]

Statutory Presumptions

Currently, some 20 states have laws creating a presumption that certain injuries arise in the line of duty. Such injuries include hernia, pneumonia, and heart attacks.[8] In general, such presumptions are rebuttable (subject to dispute)—that is, the injury is considered work-related unless proven to the contrary. The legal examination of presumption cases thus focuses upon

---

[6] Morse, "Legal Insight: Permanent Off-duty Injury," *Fire Command!*, Vol. 38, No. 2, Feb. 1971, p. 21

[7] Morse, "Legal Insight: Total Disability for a Strong Man," *Fire Command!*, Vol. 38, No. 7, July 1971, pp.28-30

[8] States currently having statutory presumptions include: Alabama, California, Connecticut, Florida, Maine, Maryland, Massachusetts, Michigan, Minnesota, Nevada, New Hampshire, New Jersey, North Dakota, Ohio, Oregon, Pennsylvania, South Carolina, Vermont, Virginia, and Wisconsin.

whether or not the "... occupational duties of a specific policeman or fireman did cause, aggravate, accelerate, or contribute to the individual's heart or lung disease, as opposed to it being simply a disease of ordinary life, which is not compensable as an occupational disease."[9]

"The effect of the statutory presumption in most states is to eliminate the requirement of proof by the claimant of employment causation. Medical evidence thus is not required of the policeman or fireman to make his *prima facie* case. In fact, with this presumption favoring the claimant, the only evidentiary facts required of record are 1. that the injured worker was employed at the moment of disability as a policeman or fireman [in a few states there is a time requirement, such as a minimum of five years' employment], and 2. that there has been a diagnosis of heart or lung disease. Once the [pension board] has received such evidence, in the employee's case, the burden shifts to the employer ... to rebut a negative, and show that the employment did not contribute to injury. An employee does not lose his presumption on vacation, on disability or in retirement."[10]

Because of their financial impact on public employee disability claims, statutory presumption laws have been under considerable challenge. Most state courts have sustained the constitutionality of these presumption statutes. In addition, courts generally hold that such statutes are reflective of a social policy, giving preferential treatment to police and fire fighters based upon a perception of the high hazards associated with their professions. As it is often difficult to establish a causal connection between a specific incident occurring in the performance of duty and the resultant injury or death, a public policy presumption is warranted.

In general, the presumption can be rebutted only when the cause of the disability is held to be non-job-related. For example, in *Fairfax County Fire and Rescue Services v. Newman*, a fire fighter was diagnosed as having pulmonary sarcoidosis, a debilitating respiratory disease of unknown cause. Medical experts for the fire fighter and for the pension board concurred that the fire fighter was afflicted and the illness would prevent his further employment as a fire fighter. Whether or not fire fighting caused the illness was highly contested by the doctors for the fire fighter and the board. In sustain-

---

[9] For example, Virginia Code Section 65.1-47.1 provides: "Presumption as to death or disability from respiratory disease. The death of, or any condition or impairment of health of, salaried or volunteer fire fighters caused by respiratory diseases, ... hypertension or heart disease, resulting in total or partial disability shall be presumed to be an occupational disease suffered in the line of duty that is covered by this act unless the contrary be shown by a preponderance of competent evidence; provided that prior to making any claim based upon such presumption, such salaried or volunteer fire fighter shall have been found free from respiratory diseases, hypertension or heart disease. . . ."

[10] Galiher, Richard W., "Presumptions for Compensation of Police and Fire Fighters," *Insurance Counsel Journal*, Jan. 1985, p. 44

ing the fire fighter's right to apply the statutory presumption and obtain a disability pension, the court noted: "Even if the negative finding . . . of no evidence of causal connection is equated *arguendo* [in the course of the argument] with an affirmative finding that there was no causal connection, the rebuttal evidence is still insufficient. We hold that to rebut the statutory presumption, the employer must adduce competent medical evidence of a non-work-related cause of the disabling disease, and there is no such evidence in the record before us."

Where a statutory presumption is not applicable (for example, in states without presumption statutes or for illnesses not covered under the statutes), evidence must show a causal connection between job incidents or responsibilities and the medical condition. In one case, a St. Louis fire fighter claimed that his disability from a spinal injury was the result of a fall while on duty. Because he felt no pain until about five months later, the pension board's medical panel ruled, at an administrative hearing, that the injury was not work-related. This ruling was upheld by the Missouri Court of Appeals, notwithstanding the plaintiff's objection that the medical records were not formally presented to him for inspection before being submitted into evidence. The court noted that at an administrative hearing it is not necessary to observe all the formalities that are customary in court. The important question was whether or not the medical records constituted substantial evidence upon which the board could find that no causal connection existed between job incidents and injuries and thus could reasonably deny the pension. Sufficient evidence was presented for the appellate court to sustain the pension board's findings that the injury was not work related.[11]

In another case, also involving the medical panel of the St. Louis Fire Department, a fire fighter claimed that his heart disease was caused by a fall while on duty. It was almost two years before he complained to his physician of discomfort. Again the Missouri Court of Appeals affirmed the pension board's finding that the fire fighter's heart disease and spinal osteoarthritis were not caused by the fall. One doctor had attributed the heart disease to several possibilities including the assertion that the fire fighter was overweight.[12]

Contagious Disease and Insanity

Numerous opportunities exist for a fire fighter to contract a contagious disease, not merely in the course of extinguishing fires but also in rescue or emergency medical activities. Considering this, the courts have justifiably held that a disease such as syphilis can be contracted in the line of duty. Such

---

[11] Morse, "Legal Insight: Presentation of Records," *Fire Command*, Vol. 39, No. 5, May 1972, p. 34

[12] Morse, "Legal Insight: Heart Ailment Not Service-Caused," *Fire Command*, Vol. 39, No. 10, Oct. 1972, p. 30

determination thus entitles a fire fighter to a disability pension if he is suffering from insanity as a result of syphilis.

Suicide

A policeman was found dead at home, a shotgun between his legs, and his head half blown off. His death was held to be "in the service of the department" within the ordinance for pensioning dependents, for there was no finding by the coroner of suicide. Also, if it can be shown that a suicide was caused by an unbroken chain of events in line of duty, suicide may be the basis for a pension claim by a fire fighter's survivors.

In *Baker v. Fire Pension Fund Commissioners*, a fire fighter's back was broken when he was riding on a fire truck that overturned. The back injury resulted in such severe and constant pain that the fire fighter committed suicide. His death was considered in the line of duty within the meaning of the pension regulations, thus permitting his widow to receive benefits. The reviewing court held: "The injuries which Baker received may justly be said to have been the proximate cause of his death. They set in motion a train of events, operating from cause to effect, that, without the intervention of any outside independent cause, resulted in his death."

Questionably Duty-Related Injury Situations

Fire fighters, like policemen, sometimes have bands. It appears that if a band player is injured by an automobile while he is running across the street to order sheet music, his injury could be considered in the performance of duty. But if he is injured in an automobile accident while on his way home to dinner, during the brief period during a 24-hour shift allowed for his meals, such injury has been held to be outside the performance of duty. In determining whether or not an injury is duty-related, the focus will be on whether a person's activities are in furtherance of the fire department's mission, are for his own convenience, or are sponsored by the department to support morale.

Recreational activities and other physical fitness undertakings frequently result in fire fighter injury. A question repeatedly occurring is whether or not such injuries are in the line of duty. To help determine if physical fitness and recreational activity are to be considered within the scope of employment, Larson[13] has developed the following three-part test. The test requires consideration of:

"1. Whether the activities occur on the employer's premises during a lunch or recreation period as a regular incident of the employment;

---

[13] Larson, Arthur, *The Law of Workers' Compensation*, Matthew Bender and Co., 1982, Section 22.00

"2. Whether the employer, by expressly or impliedly requiring participation, or by making the activity part of the services of an employee, brings the activity within the orbit of the employment; and

"3. Whether the employer derives substantial direct benefit from the activity beyond the intangible value of improvement in employee health and morale that is common to all kinds of recreation and social life."

In a case in point, *City of Tampa v. Jones*, an individual was injured while playing basketball in a city-sponsored league game. His injury was held not to be duty-related, for the employer received only tangential benefits from the league activities. Although their uniforms indicated the team's city affiliation, their activities did not directly support the mission of the city department. Even though the sporting events periodically were used to support recruitment efforts, these infrequent engagements did not support a finding that basketball play was truly "duty-related." By contrast, where fire departments encourage or require member participation in on-duty physical exercise programs, injury during such periods should be considered duty-related.

Are injuries sustained in the use of fire department vehicles duty-related? The answer depends upon whether the vehicle was being used by the individual in his employment for the city's special benefit, or for the individual's personal convenience. "While the general rule is that an employee is not acting within the scope of his employment in traveling to and from work, even though he uses a motor vehicle furnished and owned by his employer to do so, if it is shown that the employer has some special interest or derives some special benefit from his employee's use of the automobile in going to and from work, then a finding that the employee is acting within the scope of his employment is justified." (*Carreras v. McGuire*)

Some disabilities from emotional stress are caused by psychological pressures imposed by an employer. Such disabilities are considered job-connected if they stem from the employer's dissatisfaction with the employee's job performance. Thus, a stress-related disability resulting from an employer's investigation of, and dismissal of, an employee for wrongdoing has been held to be a job-connected injury. Where proof of the charges would establish misconduct outside the scope of employment, but the charges are not proved and the dismissal is set aside, any psychiatric injury resulting from the situation can be considered job-related. The alleged wrongdoing need not be within the scope of the individual's employment; the injury need only arise out of and in the course of the employment. (*Traub v. Board of Retirement*)

Injuries resulting from physical altercations between fire fighters generally are not considered "job-related." (*Meaney v. Regan*)

Delay in Presentation of Claim

In general, failure to make a timely presentation of a claim before the proper officer or commission may prevent a claimant from enforcing his right to a pension. It is essential to allege and prove that application for a pension was presented in a timely fashion and that the applicant fulfilled all the prerequisites (time in service, physical disability, contributions, etc.) in order for him to be entitled to a pension. Lack of these prerequisites may preclude judicial review of a claim.

The opposite view was expressed in an Iowa case, where a court held that a widow's right to a pension was not subject to any time limitation. Rather, the right would continue to exist all during her life and at every instance in which a pension payment was due. In a similar case, a delay of two years before presenting a claim was held to be no bar under the statute of limitations, nor under the equitable principle of *laches*. (The doctrine of laches is based on the maxim that "equity aids the vigilant and not those who slumber on their rights.")

Failure of a community to make timely pension payments can serve as an independent cause of action for negligent infliction of emotional distress. A California community entered into an $81,000 settlement with a disabled fire fighter whose pension was not promptly paid. The delays resulted in his loss of his house and in marital tensions which led to divorce.

Termination of Employment as Affecting Pension Rights

In some places, the pension statute or regulation requires a person to be a member of the department at the time of his pension application. Dismissal from the service prior to his presentation of the claim can serve to bar him from receiving a pension. Such a bar does not take into consideration the time served nor the position in which the applicant served. Thus, when an individual has resigned, has abandoned his office, or has been removed, dismissed, or discharged from office, he might not be entitled to pension benefits.

Under some plans, voluntary resignation ends all right to a pension. In the absence of a specified date, voluntary resignation is said to take effect when a copy of the resignation document is deposited in the office of the fire department or civil service commission. Subsequent withdrawal of the resignation document may not act as a reinstatement if done so in order to entitle dependents of a deceased fire fighter to a pension.

Ability to Perform Light Duty

The ability to perform light duty is entwined with the question of what constitutes "disability" as defined in a department's disability pension plan. It is universally recognized that when an individual is fully disabled and cannot perform any work, he is entitled to a disability pension. "Disability" has been defined to include inability to perform work in any position at the salary level previously held. Alternatively, "disability" has been defined as inability to perform useful and efficient service in the grade or class of position last occupied. For instance, an Illinois pension statute defines a fire fighter as a member who performs "any full-time position in the fire department." In *Pe-*

terson v. Board of Trustees, this was held to include duties in the fire prevention bureau. Thus, a person might not be considered disabled, within the meaning of the disability pension statute, only because he is unable to meet the requirements of his former position. Rather, a disability claim may require proof of inability to perform in any position at the formerly held salary or classification level. The other positions at the same level could entail lighter levels of physical exertion or "light duty," which has brought about the use of that term. (*Wells v. Police*)

Although no universal rule exists, some courts have held that a person can be required to fulfill a light duty position, at his previous classification or salary level, in lieu of the board's granting a disability pension. The reasoning behind this rule was outlined by the Oklahoma Supreme Court in the context of a police officer's application for disability allowance: "Public policy favors maximum utilization of physically handicapped employees. When the presence of physical or mental disability prevents an officer's assignment to field duty but does not inhibit his performance in other areas of police work, there can be no unfitness for duty within the meaning of the pension law. But if an officer's partial disability does in fact preclude him from serving within the department, disability allowance is indeed proper." (*Board of Trustees v. Clark*)

Thus, where there is no "useful and efficient" light-duty job, at an individual's previous position classification or salary level, that he is physically capable of performing, a disability retirement is in order. Depending upon the jurisdiction's decisional law, the employee or the department would bear the burden of showing that there are no jobs available which the employee could perform in the salary or position classification last occupied. "In short, the possibility of retirement for permanent partial disability is a function of two principal variables: 1. a permanent, partial physical impairment, and 2. the unavailability of a suitable light-duty alternative to one's former job." (*Wells v. Police*)

A similar opinion was expressed in the case of a Rockford, Illinois, fire fighter who, after injuring his back in an off-duty accident, applied for a non-service-connected disability pension. The pension was denied. Even though the man was incapable of performing the regular duties of a fire fighter, he was not incapacitated for other duties of a lighter nature, such as fire inspection work. The pension statute did not require that a fire fighter be put on disability pension unless his injuries necessitated his full retirement from service. As the fire fighter was offered an assignment as fire inspector at the same rank and pay as he had had in the fire suppression service assignment, and the individual was physically capable of performing the inspection job, denial of the disability pension was deemed justified.[14]

---

[14] Morse, "Legal Insight: No Pension for Back Injury," *Fire Command!*, Vol. 38, No. 1, Jan. 1971, p. 27

Can someone who has received a partial disability retirement, due to the lack of a light-duty position, be recalled when a light-duty job is subsequently created? A California Supreme Court decision (*Winslow v. City of Pasadena*) has affirmed a city's right to recall a disability pensioned public employee where the board had "reserved the right to monitor and reinstate those employees who had been retired on disability." Where there had been "neither guarantee nor reasonable expectation that the department would not alter its internal structure periodically, by creating, eliminating, or restructuring necessary positions or rearranging existing assignments, it was empowered to recall the retired individual." Where the city's charter permitted "reinstatement when the employee can perform the duties of the rank or position he held at the time of retirement, providing additionally that [the] reinstated member shall return at the rank and in a position of the same grade as the member occupied at the time of retirement," the city was authorized to require reinstatement.

Other Skills or Other Income

Unless a pension statute or regulation specifically makes extraneous skills or outside income a factor in fixing the amount of a pension, it is improper for a pension board to do so. However, one jurisdiction employed an "earnings test," allowing reduction of disability benefits when a retiree's outside gainful employment resulted in his total compensation exceeding the salary of current position holders. In this case, the jurisdiction was authorized to reduce pension benefits. In addition, to apply such a provision in a pension statute, the pension board may be empowered to require periodic reporting of a retiree's income, to assure his continued compliance with the prerequisites for pension benefits. Failure to comply with the filing requirement can be grounds for discontinuance of all pension benefits. *(Atchley v. City of Fresno)*

Accepting Other Employment

After receiving a retirement pension, a person's right to the pension may be suspended, by charter provision, when he reenters public employment. But in the absence of such legislation, it has been held that reemployment is no ground for termination of a retirement pension. In the fire service, there would, however, be ground for revoking a disability pension if it is established that the nature of the new employment is such that the person able to perform its duties also could perform the normal functions of a fire fighter.

Compulsory Retirement Age

The Federal Age Discrimination in Employment Act (ADEA), as amended in 1986, prohibits employers from discrimination on the basis of age against employees who are age 40 and more. Specifically, the act prohibits discharging such employees or requiring them to retire involuntarily, except when age is shown to be a "bona fide occupational qualification" (BFOQ) reasonably necessary to the normal operation of the particular business. The ADEA has been held applicable to state and local government entities and to any other employer with at least 20 workers. Thus, without a showing by the employer that age is a BFOQ for satisfactory performance of a position, age may not be considered in decisions on retirement for employees in the 40-year-and-up age group. Rather, individual assessment of an employee's fitness to perform a specific job is required.

Note that Congress has held that Federal fire fighters are subject to a mandatory retirement age (55 years) if: 1. they have sufficient years of service to qualify for a pension, and 2. their agency does not find it in the public interest to continue their employment.[15] The United States Supreme Court has sustained Congress' authority to establish a mandatory retirement age. However, extension of this authority to local government employees failed in the case of *Johnson v. Mayor and City Council of Baltimore*. As a result, state and local government fire departments must individually prove that age is a BFOQ to sustain a mandatory retirement age.[16]

Number of Fire Fighters in the U.S. by Age Group (1985)

Age	Fire Fighters	
	Number	Percent
under 20	42,150	3.9
20–29	285,600	26.5
30–39	368,600	34.2
40–49	230,650	21.4
50–59	113,200	10.5
60 and over	37,750	3.5
TOTAL	1,077,950	100.0

Source: NFPA Survey of Fire Departments for U.S. Fire Experience, 1985

Pensions as Marital Property

As a general rule, vested rights in a contributory or noncontributory pension plan are marital property if they were acquired between the date of the marriage and the commencement of a matrimonial (divorce) action. This is true even if the pension rights are unmatured at the time the action is begun. To the extent that a fire fighter has vested rights in a pension plan, they have been held to be contract rights of value, received in lieu of higher compensation, and are subject to distribution when the marriage is dissolved. This finding is based upon the monetary effect that accrual of pension rights has

[15] 5 United States Code Section 8335(b)

[16] See also *Equal Employment Opportunity Commission v. City of Altoona, Pennsylvania*. Forced retirement of older fire fighters to permit retention of newly hired members violated Age Discrimination in Employment Act.

on a couple's income. Had an individual received higher compensation in lieu of accruing pension rights, the couple either would have had greater material assets or would have enjoyed a higher standard of living. Therefore, as the delayed benefit of the pension accrued during the marital period, the pension is subject to equitable distribution between the marital partners if they divorce. Thus a court, in ordering consideration of pension rights in the distribution of marital property, can: 1. order distribution to one spouse of an equitable portion of that part of the other spouse's pension rights earned during the marriage; or 2. provide that, upon maturity of the pension rights, the recipient pay a portion of pension benefits received to the former spouse; or 3. order a distributive award in lieu of equitable distribution. (*Majauskas v. Majauskas*)

Veterans and Employees on Military Leave

The courts have ruled that, under charter or statutory provisions, time spent in military service during wars or police actions can count toward a service pension. Also, a person may be entitled to retirement benefits while on active military duty during or after release from civil service.

"Vested Right" as Applied to Pensions

Despite a vested right in a pension that was previously obtained, such right is subject to divestment under limited circumstances. For example, Indiana Code Section 36-8-6-1 provides that: "If a person who has received a benefit from the [pension] fund is convicted of a felony, is a habitual drunkard, fails to report for duty or for examination, or otherwise fails to comply with legal requirements imposed by the local board, the local board may, after notice to the person, discontinue or reduce future [pension] benefits." In upholding the constitutionality of such divestment authority, in the context of a retired police officer's felony conviction (*Ballard v. Board of Trustees*), the Illinois Supreme Court stated: "Reasonable legislative policy here provided for a method of deterring criminal acts on the part of those who might be recalled into police service. Furthermore, the legislature could have thought that payment of a pension to a felon would undermine the morale of both the general public and the active police forces of the state. It seems to us that the policeman, a public employee, took the employment subject to the conditions which the legislature saw fit to impose for reasons of public policy."

Review Activities

1. As applicable to your own jurisdiction, define what constitutes disability. What test is employed to determine whether or not a person is totally disabled?

2. Identify the name of the pension board that serves your department. What is the scope of its responsibility?
3. Can an individual be recalled from a disability retirement? If so, under what circumstances?
4. Identify the types of pension plans applicable to your department.
5. What elements are considered in calculating the amount due for a fire service pension? What elements of compensation are excluded?
6. What is the effect of filing a late pension claim? Explain.

Chapter Ten

Salary and Compensation

Federal and State Involvement

Prior to 1985, the law regulating wages and overtime provisions was developed principally on a state or local basis. This followed the United States Supreme Court holding in *National League of Cities v. Usery* that Federal regulation of these areas of particular state and local government interest (such as wage, hour, and overtime provisions for government employees) would be in violation of the Tenth Amendment to the Constitution. In 1985, the Supreme Court reversed itself and held that the Federal government *could* regulate in such areas of unique state interest. This reversal has served to initiate a critical review of public employee-management relations, the results of which are not all in at the time of this writing. Therefore, the following will generally summarize the areas of Federal government involvement in matters of salary and other compensation, with additional material from state sources that address further compensation issues.

Areas Subject to FLSA Requirements

Overtime

In the 1985 case of *Garcia v. San Antonio Metropolitan Transit Authority*, the Supreme Court held that local government employment relations are subject to provisions of the Fair Labor Standards Act (FLSA). That act sets basic minimum wage and overtime pay standards for public employees, including fire fighters. In general, it directs that each employee be paid for "all hours worked," including all time that he is required to be on duty, on the employer's premises or at a prescribed workplace of the employer. Because of the unique shift arrangements of fire fighters with their combination of duty, rest, standby, and sleep periods, special computation arrangements are provided to cover shift personnel.

Section 7(k) of the FLSA provides partial exemption from a requirement to pay a fire fighter overtime for hours worked beyond 40 hours per calendar week. Under this provision, a public employer (state or local government agency) can establish a work period of 28 or fewer consecutive days, but not less than seven consecutive days, in lieu of the standard 40-hour workweek. The work period is the basis for computing any required payment for overtime. The rule provides that where a fire fighter's work time does not exceed 212 hours in any period of 28 consecutive days, there is no requirement for payment of overtime (for time over 40 hours per week). Work periods can stretch from seven through 28 days, with each period assigned a maximum number of work hours.[1] Specific reference to the scheduling rules is advised, as any deviation from them is subject to retrospective enforcement. A two-year statute of limitations rule generally applies for recovery of back wages.

The requirement that overtime be paid, after exceeding the work hours in consecutive-day-period maximums, cannot be waived by agreement between the employer and employee. In addition, an employer cannot simply announce that overtime will not be paid at all or unless authorized in advance; he cannot escape the provisions of the FLSA this way. The employee retains the right to seek the overtime wages due. In addition, the FLSA overtime provisions are only minimum requirements; local agreements and state statutes can provide greater benefits. Note that the FLSA provides a special exception from its rules for fire departments with fewer than five full- and part-time employees. However, such departments still will be subject to state statutes and local agreements as to overtime payments.

Minimum Wage

Under provisions of the FLSA, all nonexempt employees of a state or local government must be paid not less than the current minimum wage. An exempt employee is one whose position is considered of executive, administrative, or professional quality. These positions typically are restricted to senior level people within the fire department (such as the fire chief and fire marshal). Fire fighters and line officers generally are considered to be nonexempt employees. The minimum wage rate is $3.55 an hour as of this writing.

---

[1] See *State and Local Government Employees under the Fair Labor Standards Act*, U.S. Department of Labor, WH Publication 1459, May 1985.

See also "Application of the Fair Labor Standards Act to Employees of State and Local Governments"; Final Rule, *Federal Register*, Part II, Vol. 52, No. 11, Jan. 16, 1987.

Provisions of the FLSA address only minimum wage and overtime where nonexempt employees are concerned. The subject area of salaries for exempt employees—fire chief and fire marshal—is reserved for local and state regulation. In addition, provisions for employee vacation periods, vacation pay, holidays, sick pay, and other fringe benefits remain subject to local or state control. The FLSA provides only minimum requirements; it does not prohibit negotiation for, or provision of, higher wages or better overtime benefits than those specified under the act.

Office Distinguished from Office Holder

"Office" is a legal designation for a position within local government that has specific powers and responsibilities. The office should be distinguished from an "office holder," who is the individual currently elected or appointed to carry out the duties of an office. Office holders have two principal duties. The first duty is to the council or governmental authority that created the position, specifically to fully and faithfully discharge the duties and responsibilities of the office. The second duty is a fiduciary one to the members of the general public, specifically to fully discharge the legal responsibilities of the office despite the political desires of the appointing authority. Failure to discharge either of the two principal duties can expose the office holder to personal liability.

Salary as Subject of Law or Contract

Where the fire chief and fire marshal are considered office holders not covered by a collective bargaining agreement or the FLSA, their compensation is governed by the rules applicable to municipal officers in general. No special statements can be made about office holders without reference to the specific local ordinance or charter provisions which apply. Note that without a law creating the position and authorizing the payment of salary, or without an employment contract covering salary, a fire chief cannot recover compensation for services, even after having performed them.

Vacation Time

The subject of vacation time and payment typically is covered by an employment contract or a collective bargaining agreement. Without such an agreement or a statutory requirement for the provision of vacation time (leave of absence) and payment of wages for the period of absence, there is no local obligation to provide vacation benefits. Also, the scope and extent of vacation benefits will be strictly construed and limited to the terms of the employment contract, collective bargaining agreement, or ordinance that provides the benefit.

For example, in *Chalachan v. City of Binghamton*, disabled fire fighters sued the city for payment of unused vacation time. The collective bargaining agreement provided that all city "employees" would be considered eligible for vacation compensation. However, the collective bargaining agreement was silent as to the status of fire fighters after onset of disability; it did not say whether or not they are still considered "employees" for application of the vacation compensation rule. As a result, the court was not empowered to extend the benefits of vacation payment to the disabled fire fighters. This would have been considered an unconstitutional gift of public funds. "Any additional benefits must be expressly provided for in the agreement, and petitioners' argument that they are entitled to unused vacation benefits by reason of the absence of language specifically excluding their class from vacation benefits is thus without merit," read the court's decision.

Agreement to Accept a Reduction in Salary

In general, agreement by an office holder to accept a salary less than that provided for the public office is void. Such an agreement is against public policy and will not be enforced. This is because the prescribed salary is considered the minimum necessary to obtain acceptable candidates and for satisfactory performance of the position. Thus, any salary reduction agreement would be detrimental to obtaining satisfactory candidates during any vacancy in the position. Also, the law is intended to insulate office holders, in positions of public trust, from political interference which could take the form of assaults on public officers' salaries. For example, in a case where fire fighters sought return of withheld pay, the court held: "The better rule is that where the law, as in this case, prohibits a municipality from meddling, during an officer's term of office, with the compensation he shall receive for his services, the doctrine of estoppel cannot be invoked to thwart it." In other words, the office holder will be able to recover any withheld salary.

However, a minority of courts permit salary reduction agreements made prior to initial employment but not after employment has begun. The courts distinguish this situation from the general rule because, at this point, political meddling is limited and there is no incumbent who would be affected. Rather, it would appear to be a negotiation or salary arrangement prior to employment.

Recovery of Back Pay

Where a fire fighter has been wrongly excluded from his position, he is entitled, upon reinstatement, to all his accrued salary. Depending upon the circumstances, the city may not be empowered to set-off (cancel) any earnings made by the fire fighter while working elsewhere—even in another city's fire department during the period of wrongful exclusion. (*Padden v. New York City*) Such a result was reached in a Florida case when the city of Hialeah was not permitted to set-off the $10,000 profits a fire fighter and his wife earned in their business during the period for which he was entitled to back pay.[2]

---

[2] Morse, H. Newcomb, "Legal Insight: Back Pay," *Fire Command!*, Vol. 39, No. 8, Aug. 1972, pp. 72-73

Sick Pay

In general, absence from service because of illness will not deprive an individual of compensation. Specifically, a public officer's compensation continues because the office remains filled even though the office holder is ill. For employee fire fighters, payment during periods of illness usually is covered by the collective bargaining agreement or state statute. Again, sick pay is not guaranteed under the Fair Labor Standards Act.

For example, in *Doody v. Davie* the court held that a member of the fire department, off duty as a result of stepping on a nail in his backyard, was entitled to his salary. He was considered "sick" within the charter sick leave provisions of his community.

A Michigan arbiter in 1974 found that a Pontiac fire fighter had been ordered back to duty while he was still ill. The fire fighter not only got his sick pay but was awarded an additional one-tenth of his biweekly salary. The court ruled that the fire chief could not order a fire fighter back to work without making some attempt to learn of his condition and his ability to safely perform his job.

In general, without a charter provision or ordinance there is no obligation to pay a person for unused sick leave when he retires, unless a collective bargaining agreement or other contract requires payment. However, where it has been the policy of the community to award payment for unused sick leave, the policy cannot be changed just prior to someone's imminent retirement. In *Logue v. Carthage*, the rule was changed just before a fire fighter's retirement, but enforcement of the ban on payment was prohibited. The law of equity prohibited such a change of policy because the fire fighter had relied upon availability of his unused sick leave pay during his term of employment.

Reserve or National Guard Service

Civil service regulations of the various states, as well as local collective bargaining agreements, can provide for payment of individuals during their period of Reserve or National Guard service. For example, a Clark County, Nevada, collective bargaining agreement provided: "Military Leave. Any permanent employee who is a member of the organized U.S. Army, Navy, Air Force, Coast Guard or Marine Reserves, shall be permitted up to 15 consecutive calendar days per year at annual encampment and be compensated at his/her regular pay from the county during his/her absence on any of those days which fall on his/her normal work cycle."[3]

In addition, statutes may provide for such mandatory and compensated leave. For example, New Jersey's applicable statute provides generally "... that public employees who are members of the organized militia shall be

---

[3] Agreement between Clark County Fire Fighters Local 1908 and The County of Clark, Nevada, 1982-83, Article 39

entitled to leave of absence without loss of pay or time on all days during which they shall be engaged in active duty for training or other duty ordered by the Governor."[4] Such paid leave of absence is in addition to regular vacation allowed for public employees. Note that this statutory provision for compensated leave should be distinguished from the Federally protected right of reemployment after military service. The fire fighter has not terminated his employment relationship for Reserve or National Guard duty, while he has for military service.[5]

In *Hill v. City of Camden*, a fire fighter voluntarily chose to undertake military training to further his National Guard career. After completion of voluntary training, he petitioned the city for payment of his normal salary for the period he had been away. The city's denial of his claim was affirmed by the courts. They noted that all the prerequisites for payment were present, including the governor's orders for participation in training. However, the legislative history of the statute indicated that there was no compensation for voluntarily undertaken training efforts. The legislature did not intend to give public employees the benefit of salary payments for additional, nonrequired, military training that was not essential for fulfillment of a person's minimum annual training obligation.

Assignment of Salary

In jurisdictions where fire fighters are considered office holders, they cannot assign (transfer to someone else) their unearned salaries. To do so would contravene public policy and impair the efficiency of the fire service. In addition, employer contributions to a pension fund for the benefit of one individual cannot be assigned to another person. Nor can the contributions be individually recovered in cases where employment is terminated before the individual complies with eligibility regulations such as vestment requirements for the pension plan.

Garnishment

The Federal Wage Garnishment Law sets restrictions on the amount of any employee's earnings that may be deducted in any one week or on discharge from employment by reason of garnishment. ("Garnishment" is the process whereby a person's property, money, or credits—such as wages due—are applied, according to proper statutory procedure, toward fulfillment of his debt to a third party.)

When an employee's disposable earnings—wages remaining after deductions required by law—are more than $134.00 a week, up to 25 percent of these earnings may be garnished. Where the disposable earnings are $134.00 or less, only the amount over $100.50 may be garnished.

---

[4] New Jersey Statutes Annotated 38A:4-4

[5] 38 United States Code Section 2021-2026

Where there is a garnishment order issued for the support of any person, no more than 50 percent of the disposable earnings of an individual supporting a former spouse or dependent child may be garnisheed, and no more than 60 percent where the individual is not supporting a second wife or dependent child. An additional 5 percent may be withheld in each situation if there are outstanding arrearages more than 12 weeks old.

Some state laws further limit the amounts that cannot be garnished. The Federal safeguards are considered only minimum levels of protection.

Pay Parity with Policemen

Charter or ordinance provisions of a community can provide that salaries of fire fighters be on a par with those of police officers. For example, Section 31(a) of the City of St. Louis' charter provides: "Notwithstanding any other provisions of this Charter . . . the following designated ranks and positions of members and employees of the Fire Department of the City of St. Louis shall, for the purposes of compensating said members and employees of said Fire Department by salaries for their duties, be equivalent to and correspond with the ranks and positions of officers of the police force of the City of St. Louis . . . ."

Such parity statutes usually have been sustained in court decisions. However, difficulties occur in interpreting their myriad provisions; specifically, which compensation items shall be considered subject to pay parity? St. Louis fire fighters contended that pay parity should include consideration of uniform allowances, shift differentials, and college credits. The court noted that the charter provision establishing pay parity particularly utilizes the word "salary" in lieu of the term "compensation," "salary" being defined as fixed payments made on a regular, periodic basis for the performance of regular work. Therefore, the court strictly construed the charter ordinance as limiting parity to base salary only, to the exclusion of nonregular payments. (*St. Louis Fire Fighters v. City of St. Louis*) This strict reading of the term "salary" is important, for it frequently is used in retirement plans and thus influences payments due upon a fire fighter's retirement. Reference to the discussion of retirement plan compensation, in Chapter 9, is advised.

Collective Bargaining

The predominate means for determining fire fighter salaries is through collective bargaining. For example, a 1975 Utah law permits Utah fire fighters to organize and negotiate wages, hours, and conditions of employment, and—in the event of a deadlock—to have disputes resolved by binding arbitration. In Oklahoma, in *Midwest City v. Cravens*, the supreme court upheld the state's collective bargaining and arbitration acts as constitutional; once a city agrees to collectively bargain and accept the rulings of arbiters, such rulings become binding upon the parties. Similar provisions can be found in most states' statutory enactments.

Workers' Compensation

States have passed workers' compensation or fire fighter benefit laws in order to eliminate the harsh effects on employees of their employers' common law defenses to liability. Specifically, such defenses include the rules of contributory negligence, assumption of the risk, and the fellow-servant doctrine. These laws protect employees from losses arising out of or associated with job-related injuries. Workers' compensation laws have been described as compulsory acts "... establishing in all except certain designated employments an exclusive system of compensation for injuries to employees arising out of and in the course of their employment, not caused by intoxication or intentionally self-inflicted, and resulting in disability or death. The liability so established is incident to the status of employment."[6]

In most states, fire fighters, whether volunteers or paid professionals, are considered "employees" of the city or fire department within the provisions of local workers' compensation or special fire fighter benefit laws. Thus fire fighters are protected under these laws. For example, in finding that a volunteer was an employee under the state's workers' compensation law, the Illinois Supreme Court noted: "The fact that the claimant was compensated for his services, was subject to the control of the fire chief with regard to the manner in which the work was done, was furnished tools, materials and equipment by the village, and was subject to discharge by the village board, clearly supports the finding of the employer-employee relationship." (*Village of Creve Coeur v. Industrial Commission*; see also *Office of Emergency Services v. Home Insurance Co.*)

The states are not uniform in application of the "employee" definition. In regard to Delaware's workers' compensation law,[7] a court noted: "The volunteer firemen in this case clearly do not fall under the definition of employees. There exists no contractual arrangement between [the fire company] ... and its volunteer firemen members, and the services performed by these firemen are purely voluntary and with no expectation of valuable consideration .... A volunteer rendering a gratuitous service would not come within the contemplation of the Workers' Compensation Act." (*Harleysville Mutual Insurance Company v. Five Points Fire Company*)[8]

In some other states, fire fighters come within the terms of workers' compensation because of express provisions. In other words, fire fighters, whether volunteer or paid, are defined by statute as employees under provi-

---

[6] See 5 *Southern California Law Review* 441

[7] " 'Employee' means every person in service of any corporation (private, public, municipal, or quasi-public), association, firm, or person ... under any contract of hire, express or implied, oral or written, or performing services for a valuable consideration ...." 19 Delaware Code Section 2301 (B)

[8] Action arising out of injury to fire fighters due to failure of an aerial ladder during a drill.

sions of a state's workers' compensation laws. These fire fighters are protected by law even if injured fighting a fire while off duty or outside the department's jurisdiction, such as in the mutual aid setting. Under other state statutes, anyone called upon to assist at a fire or other emergency comes within the coverage of workers' compensation. Thus, where a volunteer civilian is injured while assisting a fire fighter during a major fire, the civilian could be considered an "employee" of the city in that instance and therefore is protected under workers' compensation.

To be judged compensable under workers' compensation, an injury or illness must be occupationally related—sustained during or arising from performance of the job. Unless there is a statutory presumption that an injury arose from exposure to the work environment, proof of such exposure must be presented. For example, in *Hennige v. Fairview Fire District*, no evidence was presented that a fire lieutenant developed his anxiety neurosis due to workplace exposure. It was judged not a compensable illness under the workers' compensation law.

Workers' compensation, in general, serves the employee as the exclusive remedy for all injuries resulting from the hazardous conditions of the workplace and for negligent acts of the employer and fellow employees during the course of the work. "An employee who has a right to benefits under the Workers' Compensation Act is barred from pursuing an action against his employer for damages for personal injuries." Also, if an "... employee has a right to workers' compensation benefits on account of injury or death from injury caused by the negligence or wrong of a fellow employee, such right shall be the exclusive remedy of such injured employee and no action may be brought against such fellow employee ... unless such wrong was wilful or malicious." The reason for this limitation on remedies was stated by the Connecticut Supreme Court in *Keogh v. City of Bridgeport*: "In enacting these provisions, ... the legislature intended to avoid double liability of municipalities, which would otherwise be liable for both worker's compensation and damages for injuries caused by a negligent fireman, by preventing an injured fireman from bringing an action against his fellow employee."[9] The exclusiveness of remedy also applies to any employee's spouse or dependent who suffers a loss due to the employee's injury.

This "exclusiveness of remedy" currently is subject to significant challenge in court, but remains the general rule.[10] However, workers' compensation does not serve as the exclusive remedy of the employee where the acts of the

[9] Spouse was precluded from recovering a tort judgment against the city for the death of the fire fighter-husband, as workers' compensation benefits were held to be the exclusive remedy.

[10] See *Bernthal v. City of St. Paul*. The exclusiveness of remedy rule was held to be in violation of the equal protection clause. A municipal employee was allowed to sue the city despite his receipt of workers' compensation benefits.

employer or fellow employee are intentional, wanton, or willful. In those instances, the employee retains the independent right to sue the employer or fellow employee for injuries resulting from those grossly negligent or intentional wrongs.

Sample Cases

In *Lamalfa v. Denny*, a volunteer fire fighter directing traffic in front of the fire station was struck by the automobile of another responding volunteer. The injured fire fighter sued the fellow volunteer for his injuries. The reviewing court held that both volunteers were acting within the scope of their duties, and thus the workers' compensation benefits were the exclusive remedy for the injured fire fighter.

In *Cuna v. Board of Fire Commissioners*, the Supreme Court of New Jersey reversed a lower court decision which denied a fire fighter compensation for injuries suffered while he was sliding into home plate in a softball game sponsored by the volunteer fire department. The fire fighter was not compelled to play softball, but considered it one of his duties. The team was equipped and uniformed by the fire company, and the scores were published in the local newspaper. The court ruled that the act which compensates fire fighters for work-related injuries applied, and that a softball game, with the uniforms and publicity, was an "exhibition" which brought it under the terms of the statute. The court said that the statute allows recovery in cases where the employer as well as the employee benefits from the activity. The legislature intended liberal interpretation of recovery rules because of the humanitarian purpose of the Workers' Compensation Act.

In *Rogers v. Town of Newton*, a volunteer fire fighter with only one month's service sustained a myocardial infarction. This condition resulted in his temporary disability and his requirement for extensive and expensive medical treatment. He sued the town and its workers' compensation carrier for payment of his losses. Despite a statutory presumption in his state that heart disease arising out of fire fighting activities is work-related, the state supreme court held that the fire fighter was not entitled to workers' compensation benefits. As justification, the court showed medical evidence that a minimum of five to 15 years would be needed to develop the plaintiff's level of heart injuries; that the work done by the fire fighter on the date of injury did not involve unusual stress; and that the total of only one month's fire fighting service, involving fewer than ten fire calls, was not the actual and legal cause of the fire fighter's injuries.

In *Maines v. Cronomer Valley Fire Department*, a fire fighter was injured during hazing, part of the celebration following his induction into senior fire fighter status. The fire fighter sued the fire company for workers' compensation benefits, and sued individual fire fighters for tortious injuries. The New York Court of Appeals held that his injuries were compensable under workers' compensation as arising out of his service with the fire department. The court also found that attending meetings of the fire department was within the scope of the volunteer's employment. In addition, since the acts of the fellow fire fighters could be considered intentional wrongs—acts of assault—these fire fighters could be sued individually for all losses not covered under workers' compensation. Because an intentional act of wrong was committed, the exclusive remedy rule of workers' compensation was held not applicable in this case.

Proof of Claim

In order to prove a claimant's personal injury in a manner to justify an award of workers' compensation, it would be helpful to establish the following items by reliable evidence:

1. Good health prior to the accident.
2. Physical effects of the accident on the claimant, as related by a witness present at the time of the accident.
3. Description of claimant's normal duties, to show that he was so engaged when injured.
4. Description of his activities on the day of the accident.
5. Injury report made by his employer.
6. Medical report or testimony as to the nature, extent, and effect of the injuries, and their effect on the claimant's ability to perform his work.

Review Activities

1. Discuss with your classmates your fire department's policy on sick pay, overtime pay, recovery of back pay, and vacations. Then write a brief statement explaining how the policy conforms to the law on those subject areas, as outlined in the chapter.
2. Describe the provisions of workers' compensation. What is the purpose of the law? What are its limitations?
3. Are agreements to accept lower salary for a public office enforceable? Explain.
4. Describe your department's rules regarding leave for military service. What are the limitations of the rules?
5. Indicate whether or not your fire department is subject to a policeman-fireman pay parity rule. If so, what items are included in determining parity pay?
6. Explain "pay garnishment." What are the rules for its enforcement?

Chapter Eleven

Termination of Employment

Reductions in Force

In the absence of a state statute to the contrary, a city can create or abolish positions within its government as it finds necessary or desirable. To be considered necessary and desirable, there must be some valid, underlying reason or "just cause" for abolition of a position. Such reasons might be reductions in tax revenues or fees to support the department, reorganization of the department, or reduction in staffing requirements needed to meet desired service levels. To effect economies in municipal government is a valid ground for reducing the number of fire department personnel. This may be accomplished by abolishing positions, or by removing personnel from the payroll until further notice. Such staff reductions may be carried out even where the positions are under civil service protection, in the absence of legislative prohibition.

For example, in *Debnam v. Town of Belmont*, fire fighters were laid off when the town—despite having a substantial reserve fund—determined that tax revenues would be inadequate to support all currently authorized fire fighter positions. Fire fighters sought a writ of mandate from the courts, requiring their reinstatement. The writ was based on a state statute prohibiting fire fighter layoffs except for "just cause." In sustaining the town's authority to remove the fire fighters, the court noted: "A municipality may temporarily separate a civil service employee from his or her employment when, in the judgment of its appropriate officials, anticipated revenues will be inadequate to pay the employee's salary as well as to meet other more pressing municipal needs." In addition, a community may permanently "abolish a civil service position when, in the judgment of appropriate municipal officials, the position is no longer needed or economical." Thus, lack of money or lack of work can be "just cause" for elimination of a fire fighter position. All such determinations must be made in "good faith," must not serve as a subterfuge for removal of a personally unwanted employee, and must not serve as an avenue for discriminatory management-employee relations.

Layoffs or other reductions in force can be performed only by the agency or branch of government authorized by law. For example, the City of Utica, New York, laid off eight fire fighters recently appointed to fill positions created by the city council. The Supreme Court of New York ruled (*in Timpano v. Hanna*) that the power to create or abolish fire fighter positions rested exclusively with the city council by its own ordinances. The refusal of another branch of city government to fund the positions, after they were duly authorized by the council, was unlawful. Hence, the court ordered the authorized positions to be filled and those fire fighters who had been laid off to be reinstated.

Reductions in force or departmental reorganizations also can necessitate the non-disciplinary demotion of fire officers. Again, "just cause" must exist for demotions—for example, reductions in the size of the command group or a department reorganization for improved operational efficiency. Such demotions will be judicially sustained. In the case of *City of Erie v. Kelley*, the fire department, in response to a court order, was required to increase unit minimum manning yet was not required to increase the size (total number of personnel) of the department. The court ruled that the city was within its rights to demote officers to meet court-ordered requirements and put a new command structure into effect. Again, the city need not show a job-related connection between the conduct of the individual and the demotion. Provided the demotion is made in good faith, just cause to sustain it can exist by virtue of purely economic or operational reasons.

Removal or Suspension

There are no universal rules covering what constitutes proper procedure or sufficient grounds for the removal or suspension of a fire fighter. Minimum requirements, however, include compliance with local laws, charter provisions, civil service regulations, and rules of the fire department. The powers of removal and suspension can be exercised only by the officer, board, or tribunal designated by law, and in the manner dictated by the applicable procedure.

This rule of law was well stated in *Nelson v. Baker*, where the Oregon Supreme Court stated: "The powers of removal of an officer by the executive authorities are usually defined by the statutes upon which they depend. The executive power of removal may either be an arbitrary or a conditional one. If the power is an arbitrary one, no formalities, such as the presentment of charges or the granting of a hearing to the person removed, are necessary to its lawful exercise. A conditional or limited power of removal, as for cause, may, however, be exercised only after charges have been made against, and a hearing accorded to, the person removed, in order to effect a valid exercise of the power."

The above requirement for compliance with procedures for dismissal extends to volunteer as well as to paid fire fighters. In *Miller v. Town of Batesburg*, a volunteer fire fighter appealed his dismissal by the mayor. In sustaining the dismissal, the court was required to analyze the authority of the mayor to dismiss. The court reviewed that community's form of government to determine whether or not the mayor was vested with such authority and noted: "Under the mayor-council form of government the powers, duties, functions and responsibilities of municipal government are exercised by a council and a mayor. The mayor is the chief administrative officer of the municipality, and he is vested . . . with, among other things, the power to appoint, and when he deems it necessary for the good of the municipality, suspend or remove all municipal employees and appointive administrative officers . . . except as otherwise provided by law, or personnel rules . . . ."[1]

Removal of an Entire Fire Department

Courts have held that an entire fire department may be ousted by repealing the ordinance creating the fire fighters' positions. However, the act of placing a fire department under civil service was held not to create vacancies in the positions already filled at the time the act was adopted. Neither can a board of police and fire commissioners sidestep the rules relating to the removal of fire fighters by attempting to discharge the whole force and then immediately reappoint all but the few they intend to drop.

Some cities have entered into collective bargaining agreements which specifically exclude economic abolishment of a position as grounds for dismissal. Nevertheless, in *Schwab v. Bowen* a New York court held that such agreements cannot deprive an appointing official of the power to appoint and dismiss, provided he acts in good faith. As the city official was vested with the exclusive authority to determine the need for maintaining authorized positions, that authority could not be abolished by any contract. Therefore, termination of employees purely for economic reasons was held lawful despite provisions of the collective bargaining agreement.

Seniority as a Factor in Layoffs

As a general rule, when layoffs are necessary, the principle of seniority is followed with the last person hired the first person to be fired. Thus, where layoffs are made due to economic considerations or a fire department's reorganization, the last person employed will be the first one released. Such a

---

[1] See also *McCloud v. Whitt*. Although the city council is the only entity empowered by statute to appoint the volunteer fire chief, dismissal authority is not exclusively vested with the council. The mayor thus was empowered to discharge the volunteer fire chief.

rule is provided for in common law and usually is contained in fire fighters' collective bargaining agreements. A conflict arises where, in releasing the newly hired, the department runs counter to a judicial order to remedy illegal situations—usually in the form of sexual or minority discrimination.

In *Fire Fighters Local Union No. 1784 v. Stotts*, the United States Supreme Court addressed this issue in the context of racial discrimination, affirming the use of a seniority system's last hired-first fired rule. At the time of the layoff, the fire department was under a consent decree to eliminate the effects of prior illegal racial discrimination by hiring additional minority members. With the onset of fiscal difficulties, when the fire department was forced to lay off fire fighters, the city, in accordance with the collective bargaining agreement, applied the last hired-first fired rule. A laid-off minority fire fighter obtained an injunction, based on the consent decree, that prohibited removal of less senior minority fire fighters ahead of senior nonminority fire fighters.

The Supreme Court, in overruling the authority of the trial court to issue the injunction, held that "Title VII [of the Civil Rights Act of 1964] protects *bona fide* seniority systems, and it is inappropriate to deny an innocent employee the benefits of his seniority in order to provide a remedy in a pattern-or-practice suit such as this .... It is not an unlawful employment practice to apply different standards of compensation, or different terms, conditions, or privileges of employment pursuant to a *bona fide* seniority system, provided such differences are not the result of an intention to discriminate because of race .... If individual members of a plaintiff class demonstrate that they have been actual victims of the discriminatory practice, they may be awarded competitive seniority and given their rightful place on the seniority roster .... Even when an individual shows that the discriminatory practice has had an impact on him, he is not automatically entitled to have a nonminority employee laid off to make room for him. He may have to wait until a vacancy occurs, and if there are nonminority employees on layoff, the court must balance the equities in determining who is entitled to the job."

The Stotts case eliminated the authority of Federal district courts to order, by injunction, the reinstatement of less senior minorities over senior, nonminority fire fighters. It did not, however, eliminate the authority of a city to grant some preferences to minorities when 1. it does not entail a court order and 2. it is in voluntary response to a previously developed consent decree to eliminate prior effects of discrimination. "It is equally clear that the voluntary action available to employers and unions seeking to eradicate race discrimination may include reasonable race-conscious relief that benefits individuals who were not actual victims of discrimination." (*Local 93 v. Cleveland*)

In summary, the Supreme Court has sustained the validity of seniority systems. However, where seniority is not an issue in preferential treatment, and where the action is voluntary on the part of the union and the city, some narrowly focused preferential treatment in hiring and promotion is possible.

Terminal Leave

A growing practice in fire departments is to pay its retiring members for accumulated vacation and sick leave time that is owed them on the terminal date. Sometimes payment is in a lump sum, and sometimes on a post-retirement monthly basis; the method of payment depends on the ordinance, charter provision, statute, or collective bargaining agreement which grants the right to terminal payments. If the right was in existence at the time a member entered the fire department, its subsequent abolition prior to the fire fighter's termination of employment will not deprive him of the benefit.

For example, a member of the Roseburg, Oregon, Rural Fire Protection District had accumulated 47 days of sick leave prior to his termination. The Supreme Court of Oregon declared that the fire district's sick leave provision was part of the inducement for him to accept employment. Therefore, the provision could be considered part of the plaintiff's contract, which could not be rescinded after he had earned it.[2]

Ability to Perform Light Duty as a Bar to Removal

As previously noted in the discussion of pensions (Chapter 9), the courts are not unanimous in holding that fire fighters who are capable of performing light duty cannot be retired on a disability pension. However, in light of current financial constraints of many municipalities, placement of disabled individuals on light duty—where physically possible—in lieu of disability retirement is encouraged. (*Shelby Twp. Fire Department v. Shields*) Of greater interest is the expanding authority of fire departments to require return of previously retired disabled individuals to light duty positions. Decisions on the authority of the city to require reemployment are mixed; however, they have tended to sustain the city's authority. (See Chapter 9.)

General Grounds for Removal or Suspension

Constitutional Validity

Although the government has the right to set conditions for various aspects of conduct when a person takes public employment, such conditions cannot run afoul of an individual's constitutional rights. While a government may place conditions on employment, it may not—as a precondition—require waiver of all rights under the constitution. However, because of compelling needs of government, it may be necessary to regulate individual conduct for the public welfare—even conduct that is constitutionally protected. The courts have resolved this issue by striking a balance between the individual's

---

[2] Morse, H. Newcomb, "Legal Insight: Unused Sick Leave," *Fire Command!*, Vol. 40, No. 2, Feb. 1973, p. 29

and the government's interests regarding potentially intrusive activities. The latter are upheld only where they are necessary for some compelling governmental interest that cannot be met by any less intrusive alternative. The discussion that follows outlines how the court, in various instances, has struck the appropriate balance.

Department Rules and Regulations

A distinction must be drawn between failure of professional judgment and failure to comply with department rules and regulations governing conduct. In almost all instances, failure to comply with a department rule or regulation can serve as a basis for discipline. However, only in limited circumstances can failure of judgment, contrasted with failure to exercise any judgment, serve as a basis for discipline. To result in discipline, any failure of judgment must be significant, must be considered neglect or violation of official duty, or must be judged conduct unbecoming to a member of the fire department.[3]

The simple act of being arrested cannot serve as the sole basis for discipline. The reason was succinctly stated in *Danner v. Bristol Twp.*: "We can conceive of many instances in which a totally innocent person could be erroneously arrested without any improper motivation. It would be incorrect to hold that such an arrest could constitute grounds for suspension or removal."

A fundamental requirement of fire department rules, whether conduct requirements or elements of disciplinary procedures, is that the rules be applied in a nonarbitrary or noncapricious manner. Despite the legitimacy of the disciplinary rule, enforcement may be prohibited if the rule is enforced with discriminatory intent or in a discriminatory manner.

For example, in *Williams v. City of Montgomery*, a black fire fighter was discharged following his conviction for the crime of false pretense—a felony. Such discharge was based on the department's objective, across-the-board policy requiring discharge of any fire fighter convicted of a felony. In a subsequent discrimination suit, the court sustained the fire fighter's right to reinstatement and back pay, basing its ruling on discriminatory application of the felony rule. The court noted that there had been three earlier instances in which the personnel board had confronted the situation of a fire fighter convicted of a felony. Each case involved conviction of a nonminority fire fighter, but the board did not require dismissal in any of those earlier cases.

The court noted that the records of personnel board proceedings failed to indicate any reasoning for these deviations from the general rule. "The personnel board and the other defendants have failed to articulate the criteria by

---

[3] Appeal of *Leis*. Dismissal of a police officer was reversed where evidence showed only a failure of professional judgment. An accident following his failure to sound a warning siren, upon his entry into an intersection against the traffic light, only showed failure of judgment, as no department rule required sounding of a siren under the circumstances.

which the members of the personnel board decided that one crime necessitated dismissal while another did not; instead it appears that this determination was left to the unbridled, unguided, subjective discretion of the individual members of the personnel board.... The court feels compelled to add that there is nothing inherently unlawful, under Title VII prohibiting racial discrimination, with the defendants, including the personnel board and its members, adopting an across-the-board policy of dismissal regarding firepersons convicted of criminal offenses, as long as that policy is enforced against all firepersons, without regard to race; and further that there is nothing similarly unlawful with the defendants, including the personnel board and its members, adopting exceptions to that policy, as long as the exceptions are articulated to a reasonable degree and are made known to and applied to all firepersons, without regard to race."

Courts generally sustain a department's right to prohibit residing outside the department's jurisdiction, moonlighting, growing beards, joining subversive associations, running for political office within the service area of the department, bypassing "channels," failing to lose weight, endorsing fire protection products, refusing to attend off-duty training classes, and other such acts. Some court rulings on these constitutional questions were addressed in Chapter 5, which should be reviewed at this time. Other rulings will be addressed here—particularly rulings that involve procedural issues during any period of personnel disagreement.

A person's exercise of his Fifth Amendment right cannot serve as a basis for discipline. Recall that the Fifth Amendment prohibits forcing anyone to give testimony that could be used against him in any criminal proceeding. In a California case, *Garvin v. Chambers*, it was held that a policeman's refusal to testify against himself could not be a valid ground for his dismissal. Removal also was found improper where the sole ground was violation of a rule that no one can file a civil suit without first obtaining permission from his superior officer. Courts have stated that such a rule contravenes the constitutional principle that courts of justice shall be open to everyone. (*State ex rel. Kennedy v. Rammers; State v. Barry*)

Fire fighters have been removed without constitutional objection when the bases for removal were violation of the department's rules against engaging in direct partisan politics; being absent without leave; failure to pass the physical test for ladder work; moonlighting; and intoxication. However, "taking a drink" is not necessarily "misconduct"; for example, when it is done under doctor's orders.

A frequently cited ground for fire fighter discipline is "conduct unbecoming a member" or "detrimental to the service" of the department. Such regulations are open to reasonable, yet highly varied, interpretation. They have been held unconstitutional where they do not provide the member with sufficient notice of what conduct is required. The Supreme Court of Pennsylvania has defined "unbecoming conduct" as: "any conduct which adversely affects the morale or efficiency of the bureau to which he is assigned. It is indispensable to good government that a certain amount of discipline be maintained in the public service. Unbecoming conduct is also any conduct which has a tendency to destroy public respect for municipal employees and confidence in the operation of municipal services. It is not necessary that the alleged conduct be criminal in character nor that it be proved beyond a reasonable doubt.... It is sufficient that the complained of conduct and its attending circumstances be such as to offend publicly accepted standards of decency." (*Danner v. Bristol Twp.*)

Disciplinary punishment for unbecoming conduct has been sustained for activities that have included purchase and possession of stolen property[4] (*McDonald v. Miller*); forcible entry of a record and tape store while in an inebriated condition (*Aiello v. City of Wilmington*); witnessing yet not stopping sale of drugs by spouse within the family residence (*Travis v. City of Memphis*); and theft of a $2.97 item from a community store.

Discipline for engaging in or supporting the sale of illegal drugs should be distinguished from the mere possession of them. In a California case (*Cann v. Oakland*), a fire fighter was suspended 60 days for violating a rule against "impeding the progress, welfare, discipline, efficiency and good name of the department" when a small amount of marijuana was found in his home. Because no evidence was presented that he used the drug or that its possession had any adverse effect on the department, the fire fighter's suspension was revoked and he was awarded back pay. In *Arrastia v. Lowery*, a New York City fire fighter was denied a retirement pension when a bag of marijuana was discovered on the back seat of his car. This pension denial was reversed where there was no evidence that he owned the contraband or had knowledge of its presence in his car.

Drug Usage and Random Drug Testing

No doubt a fire department has a keen interest in discovering the extent of drug abuse by its members. In general, a municipality has the right to test for drugs if it is provided for in the collective bargaining agreement; if it occurs, following appropriate notice, during a medical screening examination for appointment to a new position or for promotion; or if it takes place during a regularly scheduled physical examination. (*Jones v. McKenzie* and *McCleod v. City of Detroit*) However, it is the *unannounced* urine test which potentially runs afoul of the constitution.

Here is a case in point: At 7:00 a.m. on May 26, 1986, the Plainfield, New Jersey, Fire Chief, assisted by the Director of Public Affairs and Safety, entered the city fire station, locked all of the station doors, awakened all of the fire fighters, and demanded that each employee submit a urine sample under

---

[4] A 15-year fire fighter veteran was discharged after an internal investigation determined that he had purchased and was in possession of merchandise stolen during multiple burglaries within the community.

the close surveillance of bonded agents employed by the city. The urine samples were to be tested for the presence of drugs. Sixteen fire fighters had positive results; they were terminated without pay. The dismissed fire fighters took their case to court, requesting that results of the testing be set aside and that the city be enjoined from further drug testing. Judge H. Lee Sarokin of the United Stated Dictrict Court in New Jersey held in favor of the fire fighters by ruling that their termination and the drug testing procedures were unconstitutional.

As the Fourth Amendment protects citizens against unreasonable search and seizure, under what circumstances can a city perform an unannounced drug test? The majority of reported cases have held that where a specific individual is identified, and facts lead to a reasonable suspicion that the individual is under the influence of drugs, the government is empowered to perform a drug test. "Reasonable suspicion" arises when an employee is observed using drugs on the job; shows significant change in behavior; is involved in unexplained serious accidents; or shows signs of being intoxicated. Where facts reasonably support a drug use suspicion, a fire fighter can be required to submit a urine sample for analysis. (*Turner v. Police*) Failure to submit a sample can be the ground for discipline. (*Krolick v. Lowery*)[5]

However, lacking reasonable suspicion that an individual is using drugs, department or station-wide random drug testing cannot constitutionally take place. By random testing, "the City of Plainfield essentially presumed the guilt of each person tested. The burden shifted to each fire fighter to submit to . . . a urine test . . . to vindicate his innocence. Such an unfounded presumption of guilt is contrary to the protections against arbitrary and intrusive government interference set forth in the Constitution . . . . Every individual has the right to be free from searches and seizures absent the establishment of some degree of reasonable suspicion against him or her." (*Capua v. Plainfield*)

The constitutional authority for random, unannounced searches for illegal drugs will be, for the forseeable future, an area of considerable controversy which undoubtedly will require resolution by the United States Supreme Court. The diversity of judgments, as outlined in the following footnote, attests to the controversy surrounding the issue.[6]

Necessity for a Fair and Impartial Hearing

Following his appointment, and potentially prior to his completion of probation, a fire fighter is considered to have a property interest in his position.

[5] A fire fighter who refused to obey an order to submit to blood testing for alcohol content, after being observed to be intoxicated, could be disciplined.

[6] *Anabel v. Ford*, Urinalysis for use of drugs is unconstitutional; *Jones v. McKenzie*, Urinalysis testing is constitutional but requires probable cause—not merely reasonable suspicion; *City of Palm Bay v. Bauman*, Urinalysis testing only requires reasonable suspicion.

Under the Due Process clause of the Fourteenth Amendment, state government and its subdivisions are required to provide at least minimum procedural safeguards, including some form of notice and hearing, before taking any action that deprives an individual of any property interest. A person's property interest in employment is created and its dimensions defined by independent sources such as state law, local ordinances, collective bargaining agreements, and employment contracts. (*Board of Regents v. Roth*) Although the existence and extent of the property interest is not defined by the United States Constitution, it does address the minimum procedural due process requirements.

Fundamental to any disciplinary proceeding is advance notice to the individual of the charges against him. Such notice must specify the charges, as well as the time of his conduct that is in question. Mere reference to the rule or regulation the individual is alleged to have violated will not constitute adequate notice. In essence, the notice must be specific and provide enough advance warning of the nature of the allegations to allow the defendant to prepare an effective defense.

To comply with constitutional due process, a hearing appropriate to the nature of the case must be provided. The timing of the hearing is based on a balance of interests: those of the city in maintaining discipline, and those of the individual in safeguarding his constitutionally protected property interest in his position. Such a hearing can be pre- or post-termination, depending on the circumstances. However, where the offense requires a post-termination hearing, the individual also must be afforded a minimal pre-termination hearing so he can respond to the charges. (*Cleveland v. Loudermill*) The pre-termination hearing need not include full rights to confront and cross-examine witnesses, or to present evidence on the employee's behalf. However, the post-termination hearing must be held within a reasonable time following the dismissal, and must provide all the constitutionally required individual protections.

Conviction for the disciplinary offense must be based on proof by a "preponderance of the evidence" of enough facts so there is a "reasonable basis" for the action. For example, if a member is charged with "conduct unbecoming" or "conduct tending to bring discredit" to the department, it is not necessary to prove facts which show violation of the penal code or which constitute a crime under any ordinance or statute. However, evidence or proof that goes beyond mere speculation needs to be shown.

In one case, a fire department captain was suspended for a month for being drunk while on duty. After the penalty was affirmed by the city council, an attempt was made to annul the council's decision. The appeal contended that the fire chief's uncorroborated testimony of the captain's intoxicated condition did not constitute sufficient evidence to support the charge. The city held that the testimony of the doctor who examined the captain, several hours after his conduct was first observed by the chief, was sufficient corroboration. In affirming the conviction and imposition of discipline, the reviewing court found that the evidence was sufficient to support the conclusion that the captain was under the influence. There was no requirement, the court said, that the chief's testimony be supported by corroborating evidence. Determinations may be based solely on the testimony of the complaining witness, providing such testimony is not shown to be incredible or impossible as a matter of law.[7]

Where the basis for discipline is a criminal offense, evidence of the criminal conviction cannot be used in the departmental hearing. This is because the parties to the disciplinary action are different, the rules of evidence are different, and a different standard of proof is involved. Similarly, a criminal acquittal will not bar a department from holding a disciplinary proceeding arising from the same set of facts. "By definition, an acquittal in a criminal case does not resolve all factual issues, but leaves factual issues doubtful.... A judgment of acquittal is only a determination that guilt has not been established beyond a reasonable doubt, although a preponderance of evidence might point thereto." (*Bredeson v. Croft*) For example, although a police officer was acquitted of embezzlement, the civil service board still was empowered to consider the factual circumstances to see if department regulations had been violated. (*City of Gadsden v. Head*)

Right to Resist Dismissal by an Improper Person

The law usually specifies who has authority to remove fire department personnel. It may be the fire chief or fire commissioner, or the right might be reserved for the civil service commission or city council.

One case arose when the city administrator of the City of Fort Smith, Arkansas, dismissed the fire chief. The chief appealed to the civil service commission, to no avail. He then appealed to the courts, contending that the city administrator had no authority to fire him; only the city's board of directors was so empowered. The Supreme Court of Arkansas held that a state statute provided that fire fighters must be given written notice of the cause of their discharge. The regulation required such notice to be given by the head or governing body of the city, so only the city's board of directors (city council) could give such notice. In this instance, neither the statute nor the regulation was complied with and the attempted discharge of the fire chief by the city administrator was held unlawful.[8]

Right to Counsel at Hearing

In establishing a hearing procedure, city charters or ordinances often provide for someone to assist the accused person in his defense. A fire depart-

---

[7] Morse, "Legal Insight: Under the Influence... Judgment," *Fire Command*, Vol. 39, No. 1, Jan. 1972, p. 30

[8] Morse, "Legal Insight: Who Fires the Chief?" *Fire Command* Vol. 40, No. 9, Sept. 1973, pp. 66-67

ment member may select any officer of the department to act as his defense counsel, or he may have a union representative or attorney assist with presentation of his defense. Thus the accused person has the right to avail himself of counsel, although the counselor need not be a member of the legal profession.

In *Wood v. City*, an ambulance driver was terminated following a hearing before the civil service commission. At the hearing, his request for a continuance, pending the arrival of his attorney, was denied. The reviewing court held that the hearing had been unfair, on the basis that effective assistance of counsel is a fundamental right of the judicial process. Likewise, a civil service commission is bound by that same fundamental principle. A disciplinary proceeding must be fair not only in fact but also in appearance. Ruling that this case was neither, the court returned the matter to the commission for a full rehearing, on the merits, with the ambulance driver's attorney present.

Right to Remain Silent

The Fifth Amendment provides that no person "shall be compelled in any criminal case to be a witness against himself." The scope of the amendment "not only protects the individual in a criminal prosecution but also privileges him not to answer official questions put to him in any other proceeding, civil or criminal, formal or informal, where the answers might incriminate him in future criminal proceedings." (*Rivera v. City of Douglas*) As noted earlier, a person retains his fundamental constitutional rights even during public employment. Thus this choice to speak out or remain silent continues during the employee's tenure.

This principle means that a fire fighter may not be discharged for failure to give his superiors an account of matters not directly related to performance of his official duties. For example, in *Cox v. City of Chattanooga*, a fire captain's employment was terminated because he refused to answer questions about a person, suspected of murder, whose name was found in his address book. The reviewing court directed his reinstatement, declaring his discharge was illegal and in violation of his constitutional right to remain silent.[9]

Garrity v. New Jersey established that, although a person is privileged to remain silent, he may be disciplined. For instance, without being required to waive immunity with respect to the use of his answers in a criminal prosecution, an employee failed to answer questions specifically, directly, and narrowly relating to the performance of his official duties. This means that a person can be subject to discipline for refusing to account for his public performance, provided the refusal would not entail the relinquishment of his constitutional right against self-incrimination. In addition, where "use immu-

---

[9] Morse, "Legal Insight: You Have the Right to Remain Silent," *Fire Command*, Vol. 42, No. 10, Oct. 1975, pp. 36-37

nity"[10] is granted for his testimony, a person's refusal to answer questions about his official conduct, on the grounds of self-incrimination, also can be grounds for discipline.

Polygraph Tests

The use of polygraph ("lie detector") tests, both for applicant background examinations and for inquiries during an employee's tenure, have had a checkered history. Initially, use of these tests for both purposes was permissible. A limited number of jurisdictions still permit such examinations. However, an expanding majority of jurisdictions, either through statute[11] or common law,[12] now restrict or completely prohibit the use of polygraph tests either for entry or for employee conduct inquiries.[13] Although the restrictions vary in scope by jurisdiction, they generally prohibit involuntary polygraph examination of current public employees. Most codes also say that refusal to submit to a voluntary polygraph examination may not serve as a basis for department discipline. (*Oberg v. Billings; Kaske v. Rockford; Long Beach City Employees Association v. Long Beach*)

Right of Judicial Review

In general, judicial review of a disciplinary action is denied on the ground that the action is ministerial in character and thus not a proper subject for review. Examples would be an employee's transfer or assignment to a punishment detail. Also, "Ordinary dismissals from government service which violate no fixed tenure or applicable statute or regulation are not subject to judicial review even if the reasons for the dismissal are alleged to be mistaken or unreasonable." (*Connick v. Myers*) However, where the discipline

---

[10] The term "use immunity" generally refers to a court order which compels a witness to give testimony of self-incriminating nature, but provides that such testimony may not be used as evidence in any subsequent prosecution of the witness.

[11] California Government Code Section 3307 provides: "No public safety officer shall be compelled to submit to a polygraph examination against his will. No disciplinary action or other recrimination shall be taken against a public safety officer refusing to submit to a polygraph examination, nor shall any comment be entered anywhere in the investigator's notes or anywhere else that the public safety officer refused to take a polygraph examination, nor shall any testimony or evidence be admissible at a subsequent hearing, trial, or proceeding, judicial or administrative, to the effect that the public safety officer refused to take a polygraph examination."

[12] States that have restricted polygraph use by judicial decision include: Florida, Illinois, Iowa, Kentucky, and Pennsylvania.

[13] States that have enacted legislation restricting the use of polygraph testing include: Alaska, California, Connecticut, Delaware, Hawaii, Idaho, Maine, Maryland, Massachusetts, Michigan, Montana, New Jersey, Oregon, Pennsylvania, Rhode Island, Washington, and Wisconsin.

imposed (such as suspension or dismissal) deprives a person of a protected property interest, or where a condition of the power to discipline is the granting of a hearing, the question of the regularity of such proceedings always is open to judicial review. This is true even though, under specific provisions of the city charter, the decision of the officer or tribunal is conclusive.

Courts generally limit a judicial inquiry to questions such as: Has there been a clear abuse of discretion by the hearing body? Was sufficient credible evidence presented to reasonably sustain the hearing body's decision? Was there compliance with all hearing requirements or other formalities? Was the charge of enough magnitude to justify the punishment imposed? and Has there been discriminatory imposition of punishment? Other matters purely of law and jurisdiction also are considered.

The court must sustain the hearing board's verdict and punishment where all hearing requirements have been complied with, where evidence is sufficient to sustain the charge, and where the punishment is neither discriminatorily imposed nor unreasonable under the circumstances. Although the court's judgment might have been different, it is powerless to substitute its own ruling for that of the hearing body. In one case, a court found that the evidence was insufficient to enable the civil service commission to verify the charges as true. The court declared that on *certiorari* (using a certified record) it had the right to examine both the findings and the evidence to determine: 1. whether the commission had jurisdiction; 2. whether it exceeded its jurisdiction; 3. whether there was any, or sufficient, evidence tending to support the charge and the findings of the hearing; and 4. whether the proceedings were conducted according to law. (*Murphy v. Houston*)

Probationary Employees

Personnel regulations generally specify that newly appointed fire fighters are subject to a probationary period, to permit direct observation of their work. The regulations typically permit probationary employees to be removed without cause during the trial period. In addition, dismissal of a probationary employee, unlike that of a permanent civil service employee, may fall within a power of summary removal. The right to notice and hearing usually is not accorded until tenured status has been attained. Sometimes the city charter, ordinances, or department rules do not set forth specific grounds for discharge of a probationer, nor authorize the civil service commission to pass judgment on the reasons for the discharge. In this situation, the reason for the discharge of a probationer is at the discretion of the person or body with appointing power. However, where personnel regulations do not provide for removal without cause, probationary fire fighters may be subject to the same rules for removal as apply to tenured employees. (*Stanfill v. City of Fairbanks*)[14]

Demotion

Once in a while, a fire fighter who has received a promotion is found by his superiors to be doing unsatisfactory work. He can be demoted without having charges preferred against him, providing it occurs during the probationary period for the higher position. A case that illustrates this principle involved a fire captain who was demoted to sergeant during his promotion probationary period. A statute permitting a public employer to reduce a probationary employee to his former rank, without giving written reasons, was challenged as being unconstitutional. The reviewing court ruled that the law was valid, saying that the essence of probationary employment is that the employer has unfettered discretion in deciding whether or not to retain a probationary employee. (*Louisville Professional Fire Fighters v. City of Louisville*) Where, however, the person being considered for demotion has completed the probationary period, he must be accorded the full procedural safeguards of any disciplinary action including notice and hearing.

Personnel Included in Removal Safeguards

It has been held that the secretary to the fire commission does not come within the scope of the protections afforded fire department personnel, with respect to the disciplining of members. Conversely, a fire department physician, veterinary surgeon, and apparatus upholsterer have been held to be entitled to the same rights accorded fire fighters.

Reinstatement

A fire fighter wrongfully suspended or removed from his position usually may seek a court order of mandamus (a command from a higher court) directing his reinstatement. "A mandamus may be directed to a public officer to compel the performance of a ministerial duty required by law," according to *Ricca v. City of Baton Rouge*.[15] Although the court can order reinstatement of an individual, it typically cannot specify the particular position or shift.

[14] This case found that removal of a probationary employee, engaged during a strike, could only be "for cause" related to his job performance. The city's personnel regulations did not permit summary removal of a nontenured employee.

[15] The court could order the reinstatement of a fire fighter, terminated in bad faith, over the objection of the fire chief and city.

Assignment of all employees remains within the sound discretion of the responsible appointing or fire department operations officer.

Review Activities

1. Explain the rules or laws of your community's fire department concerning removal or suspension of its members. In what documents are the rules found?
2. Write brief explanations for each of the following:
   A. What role does the economy play in the termination of fire department employees? Under what circumstances can termination be considered?
   B. What role does seniority play in the termination of fire department employees? Under what circumstances can an employee gain seniority other than by years of service?
3. If a fire fighter is promoted to sergeant and shortly thereafter is demoted to fire fighter, under what circumstances must there be a hearing? Under what circumstances could he appeal the decision?
4. Under what circumstances could a valid disciplinary rule be held unenforceable?
5. What items or procedures of a disciplinary hearing board is a court legally authorized to review? Can the court substitute its own judgment of guilt and penalty for that of the hearing board? Explain.
6. For disciplinary due process, what minimum requirements are necessary? Explain.

Chapter Twelve

Duty Owed by the Public to Members of the Fire Department

Owner/Occupier's Relationship to Fire Fighters

In order to determine whether or not a fire fighter has a right to recover damages from the owner or occupier of property where he has been injured, a breach of a duty owed to him must first be established. Historically, the nature and extent of any duty owed depended upon the legal status of the fire fighter entering the property: was he considered, under the law, as an invitee, licensee, or trespasser?

Today, although the majority of states still follow the historical rule that fire fighters are licensees, there are two minority positions. One holds that a property owner/occupant owes a duty to act as a "reasonable person" in all instances, considering the probability of injury to any fire fighter entering his land, without reference to his legal status. The second minority position holds that an entering fire fighter is *sui generis*—a special class of property entrant. This means that a fire fighter's right to enter the property is independent of the possessor's permission.

Because the fire fighter's right of entry is established by the presence of emergency conditions and his entry cannot be prohibited by any action or under any authority of the property owner or possessor, the fire fighter's case should be considered a special one. The history of the legal duty owed by all property owners and a description of the trespasser, licensee, and invitee rules are more fully covered in Chapter 8. However, the following will focus upon how the rules have been applied in the fire service setting.

Note that the fact that the owner of a piece of property sounded the fire alarm and thus effectively "invited" the response of the fire department does not change the legal status of the fire fighter. That status is determined as a matter of statutory or common law and does not depend upon the identity of the party who sounded the alarm or the means used to alert the fire department.

Fire Fighter as Licensee

In jurisdictions following the majority rule, a fire fighter is considered a licensee: he enters the land of someone else under a right conferred by law and in the performance of a public duty. As a licensee, he assumes the risks that may be encountered on the premises and can only recover for willful or wanton injury. Where the property owner is aware of the licensee's presence on his property, he owes a duty to exercise due care in any activities undertaken and to advise the fire fighter of any dangers associated with the owner's activities. Also, where the property owner or occupier is aware of latent (not readily visible) dangerous conditions on the property, he has a duty to warn the fire fighter of their presence. A property owner would be liable for any injuries resulting from any act of active negligence or for any "overt act of harm" to the fire fighter. As stated by the Supreme Court of Indiana in *Koehn v. Devereaux*: "A fireman is a licensee by permission of law when entering upon the property of another in performance of his duties. Therefore the fireman may recover only for a 'positive wrongful act' which results in his injuries."

The general rule was set forth in a Tennessee case, *Buckeye Cotton Oil Co. v. Campagna*, which stated: "When firemen or policemen in the course of their duties go upon the premises of an individual, the latter owes no duty to them, except to refrain from inflicting upon them a willful or wanton injury .... Firemen are authorized by law to go upon the premises of anyone in the discharge of their duties. The owner cannot prevent their entry, nor can he control their actions while they are there. They are accordingly mere licensees when acting within the city limits." For example, in *Whitten v. Miami-Dade Water and Sewer Authority*, the court held that a fire fighter could not recover for injuries suffered as a result of inhaling toxic fumes at an emergency involving the defendant's plant. As the fire fighter had the legal status of a licensee and was fully cognizant of the toxic hazard when he entered the plant, he could recover only for injuries resulting from subsequent wanton negligence or willful misconduct by plant operators.

Fire fighters are expected to know all the common hazards of fire fighting. Failure to warn them of ordinary dangers encountered at a fire—such as the possibility of a roof collapse because of the added weight of fire fighting personnel, water, and equipment—will not subject the owner to liability. In *Lunt v. Post Printing*, a Colorado court denied a fire fighter's widow damages. The "smoke" which activated the fire alarm resulted from deadly nitric acid fumes, a fact unknown then to the plant owner, operator, or fire fighters. Although the tragic result was the fire fighter's death, the responsibility for the loss could not be placed on the owner or occupant of the plant as they were unaware of the hidden danger contained in the smoke.

Some jurisdictions which follow the licensee rule have extended its application to injuries resulting from personal property fires, such as automobile fires on public roadways. "The rule, although phrased with reference to liability of owners or occupants of real property, applies to circumstances where the property, which is afire or from which the danger arises, is personal property and not located on real property owned or occupied by the owner of the personal property." (*Buchanan v. Prickett*)

Note that when fire fighters respond to a fire outside the city limits at the request of a property owner, they are under no duty to be there. Under such circumstances, in jurisdictions where fire fighters are considered licensees, the property owner owes them a higher duty of care. Under circumstances where the fire fighters are considered to be invitees, the owner owes a reasonable duty of care to make the place safe, as outlined below.

Are Fire Fighters "Invitees"?

Classification of fire fighters as licensees has not been without critical commentary. "If we accept the benefit to the land owner as the determining factor of an invitee, and it is generally so conceded, it is difficult to see on what basis, aside from *stare decisis* [standing by precedent], it can be held that a fireman or a policeman is a mere licensee."[1] Surely there can be no doubt that a fire fighter who enters a building to extinguish a fire, or to prevent spread of fire in an adjacent building, is performing a beneficial service to the owner or occupant. Under these conditions, should not the owner's or occupant's duty for the safety of the fire fighter be greater than usual?

In jurisdictions holding that a fire fighter is an invitee, the duty owed by the property owners/occupants is to "exercise reasonable care to keep their premises in safe condition." (*Luetje v. Corsini*) This duty extends to providing appropriate protection or warnings against known dangers on the property; warnings are needed where protective measures are not possible. The duty also extends to inspecting the property to see if there are any dangerous conditions. For example, in *Walsh v. Madison Park Properties Ltd.* the court held that a fire fighter, injured while inspecting a fire escape, was owed a duty of reasonable care. As the services he performed were akin to those of a building inspector or repairman, and materially benefited the property, it was reasonable to consider the inspecting fire fighter a business invitee.

In the landmark case of *Dini v. Naiditch*, a fire captain was killed and a fire fighter permanently disabled when a stairway collapsed under them during a fire. The captain's widow and the injured fire fighter sought recovery against the building owner for the death and injuries. In upholding the award of damages and overturning the licensee rule, the Illinois Supreme Court noted: "It is our opinion that since the common law rule labeling firemen as licensees is but an illogical anachronism, originating in a vastly different social order, and pock-marked by judicial refinements, it should not be perpetuated

---

[1] "Torts—Are Firemen and Policemen Licensees or Invitees?" 35 *Michigan Law Review* 1157

in the name of *stare decisis*. That doctrine does not confine our courts to the 'calf path' nor to any rule currently enjoying a numerical superiority of adherents. *Stare decisis* ought not to be the excuse for a decision where reason is lacking." The court's decision was based upon the facts that the property owners failed to provide required fire doors or fire extinguishers, permitted accumulations of trash and litter in corridors, and stored flammables under the wooden stairway which collapsed. For these reasons, the court held that the jury could conclude that the owners did not fulfill their obligation to exercise reasonable care because they failed to keep the premises in a reasonably safe condition.

Fire Fighter Considered in a Class by Himself

In some states where the duty of the property owner remains dependent upon the status of the entrant, courts have expressed difficulty in classifying the status of a fire fighter as either licensee or invitee. Because fire fighters enter premises under explicit authority of law, these courts have considered it inappropriate to put fire fighters into any classification. The reasoning is that since the fire fighter cannot be considered a trespasser—without legal authority to enter—nor a licensee or invitee, for the owner's permission to enter is not required, a fire fighter holds a "special position" with regard to an owner's property. Although the owner's duty to all others remains based upon their status, the fire fighter is owed simply a duty of reasonable care based upon the circumstances of the emergency.

Fire Fighters and the Duty of Reasonable Care

An expanding minority of states, although not adopting the *sui generis* ("in a class by himself") reasoning, have eliminated the invitee, licensee, and trespasser classification system for establishing duty. These states hold that entering fire fighters are owed a "duty of reasonable care" based upon the circumstances. Rather than focusing upon the status or classification of the entering individual, these states hold that all entrants—not just fire fighters—are owed a duty of reasonable care. The courts look at all the circumstances involved in the fire fighter's entry, such as the time, location, and type of emergency, to determine whether or not the actions of the property owner were reasonable.

The reasoning behind this departure from precedent was outlined in the case of *Rowland v. Christian*: "A man's life or limb does not become less worthy of compensation by the law nor . . . under the law because he has come upon the land of another without permission or with permission but without a business purpose. Reasonable people do not ordinarily vary their conduct depending upon such matters, and to focus upon the status of the injured party as a trespasser, licensee or invitee in order to determine the question whether the landowner has a duty of care, is contrary to our modern social mores and humanitarian values. The common law rules obscure rather than illuminate the proper considerations which should govern determination of the question of duty."

Note that no distinction is drawn between the status of a fire fighter who enters the premises to fight a fire there and a fire fighter who enters in order to gain a better position from which to fight fire in an adjoining property. The fire fighters are owed comparable duties of care.

Limitation on Liability

As noted above, there is one limitation on the liability of the property owner or occupant for failure to exercise the required duty of care. Specifically, despite any duty of care owed the fire fighter, a property owner or occupant cannot be held liable for any act or omission that directly resulted in the actual fire or emergency—the act or omission which necessitated the fire department's response. In its most simple form, the rule holds that a fire fighter who knowingly and voluntarily has confronted a hazard or risk cannot recover for injuries sustained therefrom. (*Walters v. Sloan*) The courts reason that the fire fighter knowingly and voluntarily encountered the emergency and its potential for inflicting injury. He did so by consenting to serve as a fire fighter, volunteer or paid, and by responding to the emergency; thus he knowingly assumed the risk of injury. In assuming the risk of injury, the fire fighter also should assume the risk of any losses resulting from encounters with emergencies.

In addition, as a fire fighter's compensation is based upon the inherent dangers of the occupation and he is provided with workers' compensation and other disability benefits, the courts consider it inappropriate to hold the property owner responsible for such injuries. As stated in *Luetje v. Corsini*: "Because most fires can be attributed to some form of negligence, the hazards resulting therefrom are reasonable risks assumed by those in the occupation of fighting fires and are beyond the scope of the duty owed by a landowner to remove hidden or unusual dangers on the premises or to give adequate warning thereof." This limitation, frequently called assumption of the "primary risk," is but a restatement of the "Fireman's Rule" discussed in the next section.

Rescue Doctrine

Before discussing the "Fireman's Rule," contained in the section that follows, it is important to understand another basic tenet of tort law called the "Rescue Doctrine." Under this doctrine, a tortious wrongdoer, whether his act is negligent or intentional, is liable for all injuries directly and proximately resulting from his actions. This liability includes injuries caused by or suffered by any rescuers. "The Rescue Doctrine contemplates a voluntary act by one who, in an emergency and prompted by spontaneous human motives to save human life, attempts a rescue which he had no duty to attempt by virtue of a legal obligation or duty fastened on him by his employment." (*Buchanan v. Prickett*)

On reflection, this result would not seem unjust, for the response of a rescuer is a reasonably foreseeable event. As stated by Justice Benjamin N. Cardozo in *Wagner v. International Railway*: "Danger invites rescue. The cry of distress is the summons to relief. The law does not ignore these reactions of the mind in tracing conduct to its consequences. It recognizes them as normal. It places their effects within the range of the natural and probable. The wrong that imperils life is a wrong to the imperiled victim; it is a wrong also to his rescuer. . . ."

"There is an independent duty of care owed to the rescuer himself," says one author, "which arises even when the defendant endangers no one's safety but his own. The rule is not limited to spontaneous or instinctive action, but applies even when there is time for thought. And whether the rescuer succeeds in injuring himself, or a stranger, the original wrongdoer is still liable."[2]

Fireman's Rule

Because the fire fighter's primary duty is to undertake the rescue of people exposed to danger, this question arises: to what extent, under the Rescue Doctrine, should a negligent fire starter be responsible for injuries to a fire fighter, especially in light of the fire fighter's special protections and compensation? To what extent should the wrongdoer be liable when the rescuer knowingly confronted the danger? Should the wrongdoer be relieved of liability in a case where the emergency necessitating the fire department's response had an intentional cause or reflected a dispassionate heart? Should the immunity from liability extend to all wrongful acts, including violations of building and fire safety codes? What about wrongful acts occurring after the fire department's response or after the emergency had ended?

The extent of the tortfeasor's obligation to indemnify and the various exceptions to tortfeasor liability are embodied in an appropriately named legal doctrine called the "Fireman's Rule."[3] Under the Fireman's Rule, a tortfeasor cannot be held responsible for any injuries suffered by public safety personnel if his only act of wrongdoing was the direct and proximate cause of the emergency that necessitated the police or fire department's response. As noted in *Mahoney v. Carus Chemical*, "What is meant is that it is the fireman's business to deal with that very hazard and hence, perhaps by analogy

---

[2] Prosser, William L., *Handbook of the Law of Torts*, 4th Edition, West Pub. Co., 1971, p. 277

[3] The Fireman's Rule has been widely adopted, to include the following states: Arizona, California, Colorado, Connecticut, Florida, Georgia, Illinois, Indiana, Kansas, Louisiana, Maryland, Minnesota, Missouri, Nebraska, Nevada, New Jersey, New Mexico, Ohio, Oklahoma, Pennsylvania, Rhode Island, and Tennessee.

to the contractor engaged as an expert to remedy dangerous situations, the fireman cannot complain of negligence in the creation of the very occasion for his engagement.... Probably most fires are attributable to negligence. In the final analysis, the policy decision is that it would be too burdensome to charge all who carelessly cause or fail to prevent fires with the injuries suffered by the expert retained with public funds to deal with those inevitable, although negligently created, occurrences."

This limitation also has been defined as a limitation on a fire fighter's capacity to recover from a property owner for actions for which the fire department was called. As stated in *Armstrong v. Mailand*, "Landowners owe firemen a duty of reasonable care, except to the extent firemen primarily assume the risk when entering upon the land. Firemen assume, in the primary sense, all risks reasonably apparent to them that are part of fire fighting. However, they do not assume, in a primary sense, risks that are hidden from or unanticipated by the firemen. Thus, a fireman may not recover damages from a landowner if his injury is caused by a reasonably apparent risk, But, he may recover if his injury is caused by a hidden or unanticipated risk attributable to the landowner's negligence and such negligence is the proximate cause of the injury."[4]

Thus, as in *Moreno v. Marrs*, because of the fire fighter's knowing and voluntary assumption of the risks involved in rescue efforts, the Fireman's Rule creates an exception to the general rules of liability imposed by the rescue doctrine.[5]

For example, under the Rescue Doctrine a vehicle operator who causes an accident and resulting fire would be held responsible for the following injuries: those suffered by all accident victims; those suffered by rescuers; and injuries, other than willful ones, caused by rescuers during first aid efforts. However, the situation differs where the rescuer is a fire fighter, policeman, or other emergency response person, volunteer or paid. If such a rescuer is injured in the course of rendering services in direct response to the vehicle operator's tortious act which caused the accident, the Fireman's Rule will prohibit the rescuer from recovering from the vehicle operator for any resultant injuries.

Although originally promulgated to relieve property owners from liability for fire fighter injuries occurring on real property, the Fireman's Rule has been extended to cover off-premises rescue efforts. The example above is

---

[4] See "Liability of Owner or Occupant of Premises to Fireman Coming Thereon in Dischare of His Duty," 11 *American Law Reports* 4th 583.

[5] Because the Fireman's Rule was founded upon the theory that fire fighters knowingly assume the risks involved, courts should inquire about the experience and training of the injured fire fighter to make sure that the assumption of risk was an informed one. This rule is not a universally accepted one, as most courts consider all fire fighters equivalent in training and experience and will not inquire about an individual's personal qualifications.

appropriate because it illustrates how recovery is precluded for injuries resulting from personal property and from emergencies occurring in the public domain. As the court in *Koehn v. Devereaux* reasoned: "Because it would create a dichotomy to establish policies which deny recovery to a fireman injured on-premises but allow recovery to a fireman injured off-premises, the Fireman's Rule must be applied to off-premises injuries sustained by firemen acting in their professional capacity."

Because the underlying policy reason for the rule is to limit property owner liability for injuries suffered by "professionally engaged" rescuers, application of the rule is limited to instances when the fire fighter has been called in his professional capacity. Where the fire fighter is not acting in his professional capacity, for example when he is off-duty or outside his jurisdiction, his recovery for injuries is not limited by the Fireman's Rule. Thus, there could be at the scene of an accident two equally qualified volunteer fire fighters, yet only the one rendering his services as a good samaritan and not in response to the alarm bell can avail himself of a judicial recovery for any on-scene injuries.

The Fireman's Rule has been held both applicable and inapplicable to injuries resulting from willful and wanton acts by property owners. In *Luetje v. Corsini*, fire fighters were hurt when the chimney of the defendant's building collapsed, resulting in significant personal injury. In their suit, the fire fighters alleged that their injuries resulted from the manifestly negligent or intentional acts of the property owners. The plaintiffs cited the owners' failure to keep the premises in a safe condition and to have the building comply with all applicable building and safety codes, despite repeated demands by state and village officials. The court held that the scope of the Fireman's Rule is not limited to simply negligent acts, but may extend to preclude recovery even for intentional ones. "Thus contrary to plaintiff's [assertions] . . . we do not believe that the nature of a defendant's conduct is determinative of the issue of landowner liability to an injured fire fighter."

This rule precluding recovery for acts of intentional wrongdoing, which could include arson, is not universally accepted. In a leading case, a chemical manufacturer had specific knowledge that his toxic chemical product would spontaneously combust when stored in fiber containers. He shipped the product in paper (fiber) containers, despite compelling evidence of their involvement in several fires. Even after a decision to convert to metal containers, paper containers continued to be used because they still were in stock. Within hours of a chemical shipment, in paper containers, being received at a distribution warehouse, the containers burst into flame. Results were total destruction of the warehouse structure and contents, massive evacuation of the surrounding neighborhood, and significant personal injury to responding fire fighters. (*Mahoney v. Carus Chemical*)

In exempting wanton and willful acts from application of the Fireman's Rule, the New Jersey Supreme Court noted the reasons in this way: "Hazards negligently created are staples of the duties firemen and policemen are expected to perform. Although the citizen immunity is not free from fault, the quality of fault is not so severe that the grant of immunity from liability for injuries sustained by firemen and policemen in the ordinary course of their duties offends our common sense of justice. By contrast, the degree of fault of the intentional wrongdoer is substantial. Thus, to accord immunity to one who deliberately and maliciously creates the hazard that injures the firemen or policemen stretches the policy underlying the Fireman's Rule beyond logical and justifiable limits of its principle."

The distinction between a negligent act and a willful one can be very small. This distinction may prove critical in determining whether or not an individual is subject to the Fireman's Rule exception. To be considered willful, "it must appear that the defendant with knowledge of existing conditions, and conscious from such knowledge that injury will likely or probably result from his conduct, and with reckless indifference to the consequences, consciously and intentionally does some wrongful act or omits to discharge some duty which produces the injurious result." (*McLaughlin v. Rova Farms*)

Application of the rule can immunize the property owner only from liability for the single act that necessitated the fire department's response. Prior or subsequent tortious actions of the property owner are not protected. Such actions frequently are a factor in cases where liability is premised upon a property owner's tortious misrepresentation of fire conditions, given at the time of the fire department's arrival. For example, in *Lipson v. Superior Court*, fire fighters arriving at a boil-over at a chemical plant asked the owner if any hazardous chemicals were involved or stored at the site. Despite the owner's personal knowledge that toxic chemicals were in the building and involved in the boil-over, he told the fire fighters that none was present. The fire fighters proceeded accordingly, and were injured. In sustaining the fire fighters' recovery for toxic exposure, the court stated "A fireman assumes only those hazards which are known or can reasonably be anticipated at the site of the fire."[6]

In summary, a "defendant will not be shielded from liability if (1) he fails to warn firemen of a known, but hidden danger that exists on his premises separate and apart from that to which they responded, or (2) he affirmatively misrepresents the nature of a hazard and such falsification causes their injuries." (*Rowland v. Shell Oil Co.*)[7] To apply the warning rule, the owner must

---

[6] See "Fireman's Rule—Defining Its Scope Using Cost Spreading," 71 *California Law Review* 218-252 (Jan. 1983).

[7] Fire fighters were precluded from recovery under the Fireman's Rule where, upon their arrival at a reported truck accident involving spilled gasoline, it was readily apparent that hazardous chemicals—not gasoline—were involved. As response to chemical spills traditionally is within the scope of fire fighters' ordinary duties, and the nature of this emergency was readily discernible upon their arrival, the erroneous alarm report of a gasoline spill was not a controlling factor in application of the Fireman's Rule.

have had a clear opportunity to warn the fire fighters of the danger. Where no reasonable opportunity existed or where any attempt at warning would have involved material personal danger, recovery will be precluded.

In addition to negligent or intentional misrepresentation, the Fireman's Rule also is held inapplicable where the safety officer is injured by a second negligent act or a series of them following his arrival. For example, in *Houston Belt and Terminal Railway v. O'Leary*, a fire chief was killed by the explosion of a burning box car loaded with fireworks. The Texas Appellate Court held that the defendant should have anticipated that the rough handling of the box cars could lead to a series of explosions over an extended period of time. He also should have foreseen that the fire fighters who would be called would be endangered by explosions occurring long after their arrival.

To avoid the often harsh consequences of application of the Fireman's Rule, some states have enacted statutes abolishing its enforcement. For example, by Act of May 18, 1983 (Chapter 159, 1983 Minnesota Laws 411), the Minnesota Legislature removed the common law Fireman's Rule from application within that state. Although statutory annulment of the rule is effective in eliminating its potential injustice, few states have withdrawn the rule.

Application of Statutory Safeguards to Fire Fighters

The courts are practically unanimous in holding that violation of a statute or ordinance designed to protect human life or property is *prima facie* evidence of negligence—evidence good and sufficient on its face, in the judgment of the law. In such a case, the injured party has a cause of action, provided he comes within the scope of the particular law and his injury has a direct and proximate connection with the violation. Whether or not such safeguards are intended to protect the fire fighter as well as the public is a question lacking unanimity in case law. Apparently the crucial question in each case is whether or not the legislative body, in promulgating the requirement, intended that fire fighters should come within its scope of protection.

Note that if injuries resulted from building and safety code violations, this fact will not remove a case from the scope of the Fireman's Rule, discussed above. As courts have stated: "We see no reason for a different result where the conduct that causes the fire is also a violation of a statute or ordinance. As stated previously, the function of a fireman is to deal with fires and he assumes the risks normally associated with that function." (*Washington v. Atlantic Richfield*) However, where the statute is specifically promulgated to protect fire fighters, it usually will not be subject to the limitations of the Fireman's Rule.

Application of NFPA Standards to Fire Fighters

Even if there is no ordinance specifying installation of safety devices to prevent a fire or explosion, a landowner or factory operator may be found liable. The basis can be failing to adhere to nationally recognized standards considered to be "good practice within the community," such as those promulgated by the National Fire Protection Association (NFPA).

For example, in *Bartels v. Continental Oil* an action was brought for the wrongful death of a fire captain. He died as a result of injuries suffered when a gasoline tank exploded while he was fighting a fire in the defendant's bulk storage plant and filling station. A kerosene tank had already ruptured, providing additional fuel for the flames issuing from a tank truck that had been ignited when the truck's driver tried to demonstrate his new cigarette lighter. Within the next half hour, two more storage tanks ruptured. Shortly after that, the fourth tank left its concrete cradle and "rocketed" about 100 feet over the filling station; its 15,000 gallons of gasoline formed a ball of fire that killed one bystander and injured five fire fighters and 23 civilians.

No ordinance required retrofitting of larger tank vents. However, the oil company's safety director, because of his knowledge of national standards and his professional position, knew that the existing vents were not of sufficient capacity. Also, because plant management knew that larger emergency vents would be safer, the Supreme Court of Missouri found that the existing undersized vents created a hidden danger. Although known to the defendant, existence of the hazard was unknown to the fire fighters. Further, the oil company could reasonably have expected that such a tank might rocket, yet it failed to warn the fire fighters of this risk. These hidden dangers, although not code violations, should have been disclosed, the court ruled.

Duty of a City as a Property Owner

Local governments, as property owners, have a duty of care with respect to the maintenance, care, and safe conditions of their property. This duty is similar to the duty of other property owners in the community which was outlined in detail in Chapter 8 and reviewed above. The duty of fire departments to fire fighters, to maintain safe facilities and equipment (as outlined in Chapter 7), typically involves issues of fire fighter occupational and workplace safety rather than premises liability. However, when the fire department inspects or responds to a call to city-owned property other than fire department facilities, the duty of the city to its fire fighters is similar to that of the public to members of the fire department. The city will be bound by the rules previously discussed for licensee/invitee, *sui generis*, or reasonable duty of care, as appropriate, for the specific jurisdiction.[8]

---

[8] For injuries sustained in a fall through a roof, based upon the theory of failure to keep a city-owned building in good repair, the city was held liable to a fire fighter for a $2.6 million judgment in *Donnelly v. City of New York*.

Duty Owed by Utility Companies

Apart from the landowner-fire fighter relationship, another consideration involves the duty owed by power and gas companies to safeguard their equipment to prevent injury to fire fighters. As the utility can be engaged in the delivery of potentially hazardous commodities—electricity, the various forms of flammable and explosive gas, high pressure steam, etc.—some questions arise: What duty of care is owed by the utility? Is the utility governed by the duty of care owed by the owner of the property where the emergency occurred? What duty of care is owed by the utility after the emergency?

In general, a utility has the duty to keep all of its dangerous equipment and other holdings in a reasonably safe condition. Where statutory or regulatory requirements include specific precautions for transmission of its hazardous materials, the utility is obligated to comply within a reasonable time. Once dangerous conditions have been uncovered and the utility notified of their existence, the utility is obligated to respond promptly to eliminate the hazard. It also is required, in accordance with custom and practice of the industry, to inspect its facilities for potentially hazardous conditions.

For example, in *Verges v. New Orleans Public Service*, the electrical utility was sued by a fire fighter. He was injured when the boom of the fire department's "Squirt Unit" came in contact with overhead power lines during a training exercise. Court testimony indicated that placement of the 8,000-volt electrical conductors in front of the fire station—the site of the training exercise—was not fully in compliance with local code. Although the utility was aware that the location periodically was used for training, no relocation of the wires was performed. In denying recovery, the court noted that a suitable training location, away from all electrical transmission wires, existed at the rear of the station. Also, it noted that placement of the wiring was typical of that found throughout the city and of that which city fire fighters encountered on a recurring basis. As evidence showed that the boom of the Squirt Unit had come in contact with the wires, recovery was prohibited. The court ruled that the utility had acted reasonably under the circumstances.

In a California case, a Fresno fire fighter was electrocuted by a wire which had fallen at a fire. The fire department had not asked to have the power turned off when the wire was discovered, but an agent of the power company, who happened to be at the fire as a spectator, knew that the line was down. The court held that there was no negligence in the power company's failure to have an employee on duty at the fire because there was no ordinance requiring it, nor was there a request by the fire department for the utility to respond. (*Pennebacker v. San Joaquin Light and Power*)

Summary of Public's Duty Owed to Fire Fighters

1. Duty of Owner-Occupant:
   A. In all instances, to avoid wanton or willful misconduct and active negligence.
   B. In all instances, to warn fire fighters of hidden dangers—any "trap" about which the possessor knows and has a reasonable and safe opportunity to communicate.
   C. In all instances, to maintain in a safe condition the ordinary means of access to the premises, as well as the fire escapes, smoke towers, or other safety features required by law.
   D. In all instances, to allow fire fighters access to all portions of the building, whether to fight fire in the structure or in adjoining property.
   E. jurisdictions which consider fire fighters invitees *sui generis*, or due a duty of reasonable care, to maintain property in a reasonable condition and to inspect the premises to assure the safety of entrants.
2. Duty of Utility Companies
   A. To maintain their dangerous equipment and processes in a reasonably safe condition.
   B. To repair or control dangerous conditions which could lead to the injury of fire fighters, where:
      a. Notified by the fire department or other party, or when
      b. Required by law to send a repairman upon receiving notification of a fire or other danger.
3. Duty of Motorists
   By statute or ordinance, motorists generally are forbidden, among other things, to follow too closely behind fire trucks and drive over fire hoses, and must yield the right of way to fire apparatus displaying the appropriate audio and visual warning signals.

4. Duty of All Citizens
   A. Do not interfere with the fire department's performance of emergency services.
   B. Leave the emergency scene when so directed.
   C. Where required by state law, assist fire fighting forces when requested to do so.

Review Activities

1. Describe how, in an area with no ordinance or code requirement specifying safety devices to prevent a fire or explosion, a property owner or occupant may be found liable for injuries to fire fighters.

2. In your own words, write a definition of the so-called "Fireman's Rule" that is applicable to your jurisdiction. Give examples of situations in which fire fighters would and would not recover for injuries.
3. When fire fighters are classified as licensees, what is the duty of care owed by the property owner or occupant? When could the property owner be found guilty of negligence?
4. Describe the underlying policy reasons behind the "Fireman's Rule." Do you think they are still valid today with the ever-increasing understanding of the hazards involved in the fire fighting profession? Explain.
5. Write a description of your state's statutory safeguards for the protection of fire fighters.
6. What is the duty of your state's utility companies in the event of a fire emergency? Do they have an obligation to respond? If so, under what circumstances and at whose directive?

Chapter Thirteen

Liabilities of Fire Fighters

In General

The general liability rule, by both court decision and statute, is that every person is bound to conduct himself as a reasonable person would, considering the circumstances, without causing injury to the person or property of someone else. A person may be liable for injuries inflicted unintentionally, as well as for those caused deliberately, where his conduct fails to come up to the standard of care exercised by the ordinary, reasonable, and prudent individual. An act of negligence can occur in any of three ways: through misfeasance, malfeasance, or nonfeasance of the wrongdoer. Misfeasance is the improper doing of an act which a person might lawfully do; malfeasance is the doing of an act which a person ought not to do; and nonfeasance is the failure to fulfill an obligation.

Under certain circumstances, a person may be liable for damage sustained by others even though he has done everything humanly possible to prevent the injuries. This rule of strict liability is more fully outlined in Chapter 8 but is worth repeating here. Strict liability extends to those activities which are so fraught with danger, or which present such extreme risk of harm to others, that public policy demands that the individual be responsible for all results without consideration of his conduct.

In earlier chapters, the rules of negligence and strict liability generally have been discussed with the city being considered the defendant. Cities are the principal defendants in all tort actions against cities for their activities. However, as the public employee or officer who perpetrated the alleged tortious offense also can be held liable, discussion of potential personal liability is appropriate. In the sections that follow, the liability exposure of individual fire fighters is explored. Although fire fighters, like other public officials, are frequently indemnified for their liability and defense costs by their employing community, this does not help define their total liability exposure. Also, where statutory or constitutional law prohibits indemnification for specific types of judgments, such as punitive damages, fire fighters remain subject to that liability exposure. Because the scope of the city's and the individual's respective liability exposures is not identical, individual liability warrants separate discussion.

Fire Fighters as Public Officers

There is no unanimity of judicial opinion on the question of whether fire fighters are "public officers" or mere public "employees." In those jurisdictions favoring the public officer designation, fire fighters are subject to the duties of care imposed upon them as individuals, as well as to those duties accompanying their status as public officers. The designation of "public officer" usually is made by ordinance or statute. On the other hand, where fire fighters are considered to be employees, they are subject only to liability for failure to perform their individual duties.

Breadth of Fire Fighters' and Officers' "Scope of Authority"

It generally is conceded that the operation of government would be severly handicapped if officials were to be held answerable in damages for mistakes, negligence, or poor judgment in the honest performance of their duties. The very function and duty of officials requires them to make decisions involving judgment and discretion. Fire officers and fire fighters often must make discretionary decisions; for example, to tear down a building in the path of a fire, to raze a fence to gain access to the rear yard of a building adjacent to one on fire, or to issue a hazard citation. Courts tend not to "Monday Morning Quarterback" the fire officer, inspector, and fire fighter as to what activities should be performed or which actions are within their authority. In essence, the issue of what involves fire safety, within the scope of the fire department's authority, is considered very broadly. However, this breadth of authority does not extend to the deprivation of an individual's constitutional rights.

Civil Rights Violations

A city, public officer, or employee depriving an individual—under color of state or local law—of any constitutionally protected right (that is, life, liberty, or property interest) is liable in damages to the injured party. Specifically, Title 42 United States Code Section 1983 states: "Every person who, under color of any statute, ordinance, regulation, custom or usage, of any State or Territory or the District of Columbia, subjects or causes to be subjected, any citizen of the United States or other person within the jurisdiction thereof to the deprivation of any rights, privileges or immunities secured by the Constitution and laws, shall be liable to the party in an action at law, suit in equity, or other proper proceeding for redress."

In the interim since the First Edition of this text, Section 1983 actions have been the area of most controversy within local government circles, as well as an area of increasing liability exposure of fire fighters, officers, and fire departments.

Although very similar to a normal negligence cause of action (such as showing that a breach of duty proximately caused a plaintiff's injury), the elements of the Section 1983 cause of action are slightly different. Section 1983 requires:

1. Existence of a Federally secured interest based upon the United States Constitution or Federal statute.
2. A deprivation of the interest under apparent terms of state or local law. This deprivation is not limited to actions of public officers or employees, but can be by an individual having some governmental involvement or connection.
3. Injury proximately resulting from the deprivation. In most cases, only the actual injury need be proven. Proof of intent to injure usually is not required.
4. Damages to the plaintiff.

Section 1983 actions have arisen in various aspects of fire department operations. These actions pertain to all aspects of fire department services, from actions alleging intentional discrimination in fire fighter hiring and promotion (outlined in Chapter 5), to discriminatory performance of fire suppression services (outlined in Chapter 6) and deprivation of property interests (outlined in Chapter 8). Regarding the liability exposure of the fire officer and fire fighter, it is important to note that Section 1983 covers activities of individuals in both their official and individual capacities. Finally, state immunity statutes generally do not apply to protecting either the city or the individual from liability resulting from a Section 1983 action.[1]

A case in point is *Watson v. McGee*, alleging deprivation of a constitutionally protected liberty—interest in personal safety. When people held in jail were injured in a fire, liability for the negligent maintenance of non-fire safe conditions could be assessed against individual city officials, the court held.

---

[1] As stated in *Owen v. City of Independence*, "The municipality's 'governmental immunity' is obviously abrogated by the sovereign's enactment of a statute making it amenable to suit. Section 1983 was just such a statute. By including municipalities within the class of 'persons' subject to liability for violations of the Federal Constitution and laws, Congress—the supreme sovereign on matters of Federal law—abolished whatever immunity the municipality possessed."

This liability potentially could include the fire chief and fire inspector where they have any obligation to inspect jail facilities and seek corrective measures.

In another case, *Wheeler v. City of Pleasant Grove*, an apartment builder obtained all required building approvals. However, when a subsequent zoning ordinance (later held invalid) was passed and this action resulted in revocation of the building permit, a Section 1983 cause of action existed. The effect of the subsequent ordinance passage was considered confiscatory of the builder's vested property interest in proceeding with his building effort. The city having granted all approvals, thus vesting the builder's rights, the subsequent withdrawal of approval became an unconstitutional taking of property without just compensation.

Where a fire department policy would provide a discriminatory emergency response pattern, the city, fire chief, and dispatchers could be held liable for deprivation of a constitutionally protected property or liberty interest. (*Archie v. City of Racine*) Section 1983 is intended to eliminate any policy, custom, or practice that results in a deprivation of a life, liberty, or property interest. Cities can be held liable for existence and exercise of such policies. Therefore, even one civil rights violation involving an abuse of official power by a city officer or official has been held to be a Section 1983 compensable injury. (*Pembaur v. City of Cincinnati*) Further discussion of Section 1983 actions in Chapters 5, 6, and 8 should be reviewed at this time.

Liability as a Public Officer

In some situations, the public officer's delegated power is strictly construed. Specifically, if a fire fighter or officer causes legal detriment to an individual under certain circumstances, he could be subject to civil liability in damages. The circumstances are: 1. Acting Beyond Scope of Authoriy, and 2. No Authority to Act.

1. *Acting Beyond Scope of Authority*. For example, the fire officer's duties may be limited by law to the extinguishment and prevention of fires. Directives issued by the officer to contain a noncombustible, toxic gas leak could be considered beyond the scope of his authority, the occurrence of a fire incident being physically impossible. However, the authority of fire officers under such circumstances is liberally construed by the courts to include authority under almost all hazardous conditions, even those where fire cannot occur.

Examples of situations where liability could be imposed for acting outside the scope of authority include the following:

HazMat Activities

In the absence of legal authority to operate hazardous materials teams, the response to spills, leaks, and other non-fire emergencies would be *ultra vires* (beyond their official authority). The responding agencies thus would be liable for negligent acts in coping with such incidents.

Riot Suppression

Where the authority of the fire department is limited to extinguishing and preventing uncontrolled fires, and enforcing fire codes, fire fighters cannot lawfully use their hose streams to assist the police in quelling riots. Such an action clearly would be considered *ultra vires* and would render the fire fighters personally liable for any injuries caused by such tactics. However, if fire fighters engaged in fighting a fire were attacked by rioters, any reasonable self defense would be justified. This could include use of deadly force in case of imminent, personal exposure to death or serious bodily harm to the fire fighters. However, no circumstances would warrant use of excessive force or deadly force to arrest rioters. Under such circumstances, fire fighters should call for law enforcement officers to provide protection. If no police are available, the fire fighters should withdraw to a position of safety, and continue their suppression activities only when the situation improves. If retention of their position is necessary to perform rescue work, then only the force needed to assure individual safety and to carry out rescues would be justified.

Rescue Work

Undoubtedly, the saving of human life, as incident to fire extinguishment, is clearly within the province of a fire fighter's duties. Yet when a fire fighter responds to accident scenes or to private homes to give medical assistance, he could be acting in an *ultra vires* capacity unless there is legal authority for providing such service. A fire fighter's humanitarian motives may not be a bar to liability if, through failure to exercise reasonable care, he injures the person he is treating. Clearly, where the fire department has charter authorization to operate an Emergency Medical Service (EMS), or the state constitution or a state statute empowers cities and departments to perform EMS activities, such efforts would be within a fire fighter's scope of authority. Reference to the applicable enabling rules is necessary to fully determine a person's "scope of authority."

A Seattle, Washington lawsuit involved two fire fighters who responded to an emergency call when a child was locked in a bathroom. The case raised the question of whether or not they were acting within the scope of their authority and hence if they had the right to use their emergency warning devices. If their activities were unauthorized, their act of exceeding the rules of the road (that is, speeding and traveling through a red light with siren and warning lights working) could be considered *negligence per se*, and would subject both the fire department and fire fighters to liability.

About three-fourths of the states have "Good Samaritan" laws. These laws provide that someone who voluntarily tries to help an injured person shall not be chargeable with any fault nor be legally responsible for any errors or omissions in the care that he gives. Although the laws vary by state, their general effect is that the individual is immune from liability for acts of negligence. However, that immunity is not absolute. Most states provide that the immunity from liability shall not apply in case of gross negligence (such as a person's reckless disregard for the consequences of his actions) and willful and wanton misconduct (that is, intentionally injurious acts).

Some states have extended the "Good Samaritan" coverage to paramedics. For example, California state law provides that a paramedic who follows the instructions of a doctor or nurse will be immune from civil liability to the same extent as physicians and nurses, now protected when they give voluntary emergency services.

Salvage Work

Under the power to extinguish fires and save lives, fire fighters responding to an alarm of fire need have no apprehensions about activities designed to prevent water damage to the property involved. This is true whether the water issues from hose streams or automatic sprinkler systems. However, when the call is in no way related to a fire—as where fire fighters are called to pump out a cellar flooded by a broken pipe, or to patch up a homeowner's roof in a rain storm—their endeavors potentially take them beyond their scope of authority. Again, reference to the specific powers of the fire department, under non-fire conditions, is warranted.

Demolition Squads

Some fire departments have a group of individuals who are trained to handle high explosives, so that if it becomes necessary to blast fire breaks in order to stop fire spread, qualified personnel are available to do this work. The fire department's authority to dynamite buildings in the path of a fire is no longer questioned, such authority being part of the inherent power of the fire chief to contain the spread of fire. Sometimes, however, a person is injured due to the negligent use of explosives while fire fighters are doing such things as helping road maintenance crews to remove old bridge footings. Then liabiilty for this hazardous activity cannot be averted on the ground that it was done in the course of a fire fighter's duty. A fire fighter's duty does not include bridge maintenance and repair.

2. *No Authority to Act.* The authority to act in a given field may not exist or may not be vested in a given officer. For example, if this authority under the law is limited to the extinguishment and prevention of fires, a fire fighter has no authority to tell an individual how to construct his building. This authority typically lies with the building official. The fire

fighter can only inform the builder that, in the event the stated activity is conducted, specified fire prevention items, as outlined by the Fire Prevention Code, are required.

Lack of Care in Exercising Power

Although a person has the authority to act and invokes such authority in an appropriate fashion, liability still can result if the authority is exercised without due care. For example, a fire fighter needlessly throws antique items of value—not exposed to threat of fire—out of a second story window. Such an effort to save the owner's property might be considered failure to exercise reasonable care. It is of little consolation to the owner of the antiques to say that a fire fighter was "saving" his property when the remnants of his centuries-old artifacts are strewn all over the sidewalk.

First Aid Volunteers (EMS)

The general rule for all providers of first aid, whether specially trained or not, is that they have the duty to exercise reasonable care under the circumstances. Often an individual, because of his professional position or employment, is required to possess special skills or certification. Such a person will be held to the duty of care that would be exercised by individuals in good standing of comparable skill or certification. For the Emergency Medical Technician (EMT), this means that he is obliged to exercise the same degree of skill, care, and judgment as a comparable first aid provider, with similar skill, experience, and certification, working under the same circumstances. Without statutory immunity or a fire department indemnification provision against personal liability, the first aid provider would be personally liable for a plaintiff's injuries. This would occur in failure to provide a reasonable level of care, which directly and proximately resulted in the injuries.

Throughout discussion of negligence in medical care, it is important to remember that the negligent act must cause, or make worse, the patient's injuries for there to be liability. Where a negligent act does not cause injury there is no resulting liability. Again two factors—negligence and injury—must be shown for there to be liability. For example, in *Miller v. Mountain Valley Ambulance Service*, the defendant ambulance service responded to the scene of a construction accident where a man had suffered a severe electrical shock. The victim was transported to the hospital, where he was dead on arrival (DOA). Apparently, during the transport the patient aspirated his gastric contents and asphyxiated on them. Suit was brought against the city and the ambulance service for negligence in providing care. In affirming the jury's verdict for the ambulance service, the court noted that: "Negligence results in liability only if accompanied by causation." Thus, where the cause of the injuries was apparently the electric shock and not the alleged negligent medical care, the ambulance service and EMTs could not be held liable.

Because of the inherent risks associated with emergency medical care, many states have passed first aid immunity statutes to cover fire fighters and EMTs. For example, the General Laws of Massachusetts, Chapter 111C, Section 14, provides: "No emergency medical technician certified under the provisions of this chapter and no police officer or fire fighter, who in the performance of his duties and in good faith renders emergency first aid or transportation to an injured person or to a person incapacitated by illness, shall be *personally* in any way liable as a result of rendering such aid or as a result of transporting such person to a hospital or other safe place, nor shall he be liable to a hospital for its expenses if, under emergency conditions, he causes the admission of such person to said hospital."

As noted earlier, the scope of an immunity statute can protect the individual fire fighter or EMT, yet not protect the employing fire department or community. For example, in *Taplin v. Town of Chatham* two certified EMTs of the fire department responded to an injury scene. After ascertaining the patient's condition, they transported him to a local home instead of a hospital. The following day, the patient was admitted to a hospital where he underwent emergency brain surgery and subsequently died of his injuries. Suit was brought against the EMTs and the town. In holding the EMT immune and the town liable for the death, the court noted that the immunity statute does not extend protections to the employing fire department or the town. The legislative history of the statute did not indicate any intent to protect a town or city.[2] Likewise, the common law rule of *respondeat superior* ("Let the master answer") extends only liability, not immunity, to the employer.[3]

In contrast with specific functional immunity (that is, immunity for performance of a specific public function), statutory immunity can be provided to communities for the provision of all governmental services. Emergency medical service may be considered but one of the governmental functions. In addition, where the service is provided by the fire department, the EMTs and paramedics of the department could be considered "fire fighters" as defined by the immunity statute and thus be personally protected from liability. For example, in *King v. Williams* an ambulance of the City of Akron, Ohio, while on an emergency response, entered an intersection against the traffic signal

[2] In refusing to sign and returning a prior immunity bill to the General Court (legislature) of Massachusetts, the Governor stated: "The bill should clearly indicate that the exemption from liability applies only to the individual policeman, fireman, or technician, and not to his employer, so that injured persons will not be denied any recourse against any other party that would otherwise be available."

[3] See also *Raynor v. Arcata*. Statutory immunity of a municipal employee does not free a city from liability.

and struck a passenger vehicle. The injured couple in the car sued the city and the driver for personal injuries resulting from the accident. In holding the city and ambulance driver immune from liability, the court noted that Ohio's applicable immunity statute for governmental functions can be interpreted to include emergency medical services performed by city fire departments.[4] Therefore, despite the plaintiffs' injuries, recovery against the city and driver was prohibited. In essence, they possessed a complete defense against any claim of negligence in providing emergency medical services.

Standard of Care in Rescue Service

Again, in the absence of an immunity statute a fire fighter or EMT can be held liable for negligent acts. As a general rule, the fire fighter's duty is to carry out those rescue efforts for which he is trained, and in the manner in which he has been trained.

Failure to Respond

As discussed in Chapter 6 under Emergency Dispatch Services, failure to respond when alerted to an emergency situation can serve as a basis for liability. Again, without statutory or common law immunity, where a special relationship is found to exist, failure to respond to an emergency medical call can be negligence. In *Archie v. City of Racine*, the fire department's dispatcher received multiple calls from a city's "character," a habitual drinker, reporting that a female friend was breathing fast and in need of medical assistance. The dispatch tapes clearly recorded the heavy breathing and stress of the friend, as well as the anxiety of the "character"—despite his slurred speech—in reporting her condition. Contrary to a department policy to respond to "all emergency calls," the dispatcher advised the woman to breathe into a bag to slow her respiration rate, but did not send the rescue unit to verify her condition. After several more calls, but no ambulance response, the woman died. Although the court could not hold the dispatcher liable because of any civil rights violation, it stated that the dispatcher's judgment was flawed and could be considered negligent under common law terms. Even though the "character" might have been "under the influence," he could clearly understand the urgency of the situation and the need for medical attention. The failure to dispatch the ambulance was undoubtedly wrong.

---

[4] Ohio Revised Code Section 701.02 provides: "The defense that the officer, agent, or servant of the municipal corporation was engaged in performing a governmental function, shall be a full defense as to the negligence of: . . ."(B) Members of the fire department while engaged in duty at a fire, or while proceeding toward a place where a fire is in progress or is believed in progress, or in answering any other emergency alarm. . . . Firemen shall not be personally liable for damages for injury or loss to persons or property and for death caused while engaged in the operation of a motor vehicle in the performance of a governmental function."

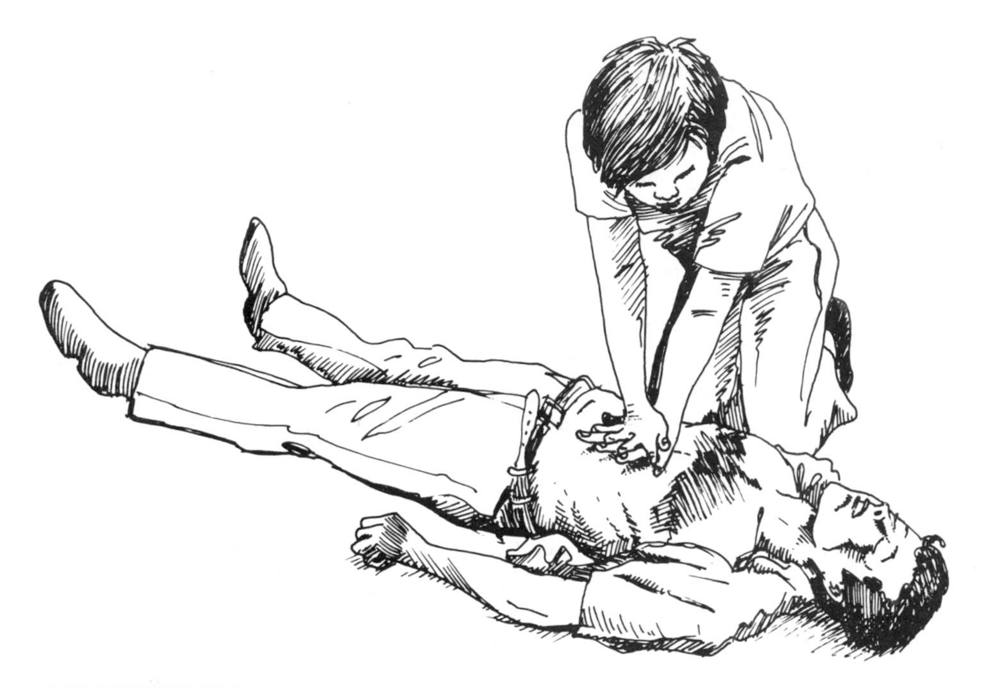

SOME FIRE DEPARTMENT MEMBERS are Emergency Medical Technicians, trained to perform life-saving services at the fire scene. Cardiopulmonary resuscitation (CPR) might be necessary to revive a victim rescued from a smokey environment.

Cardiopulmonary Resuscitation

Practically all fire fighters today are trained in standard first aid practices including cardiopulmonary resuscitation (CPR). In a situation where this procedure was called for in the presence of a fire fighter or EMT, failure to at least attempt it might be considered negligence. However, negligence is not limited to failure to perform acts called for by CPR training; it also can result from failure to carry out the required techniques in a proper manner. For example, Dr. William H. L. Dornette has cautioned that "pushing sideways instead of straight down when performing CPR may fracture ribs and puncture lungs. This misapplication of the technique can well be considered to be a negligent act, even though it is possible to break ribs if the technique is applied properly." After pointing out that a fire fighter does not have to possess the skill of a physician to attempt resuscitation, Dr. Dornette stated that, however, the EMT would be liable if he did not demonstrate the degree of skill and care possessed by comparably trained and equipped rescue personnel. Moreover, the emergency nature of the situation is always taken into consideration in assessing the standard of care.[5]

[5] Dornette, William H. L., "Fire Fighters, Resuscitation and the Law," *Fire Command!*, Vol. 38, No. 7, July 1971, pp. 13-15

Transporting the Patient

Failure to transport a patient, or transporting him to a residence instead of a hospital, can serve as a basis for finding negligence. For example, in the *Taplin* case noted earlier, the basis for potential EMT liability was failure to transport the patient to a medical facility where a complete medical examination could be performed by a licensed physician. Apparently, the EMTs assumed that, based upon his physical condition, the patient did not need hospital care. This was a questionable decision more appropriately made by a licensed physician. In summary, where there is any question or doubt, it is appropriate to transport the patient, or arrange for transport, to a medical facility.

Care Required in Physical Rescues

Not everyone is grateful for being rescued. For example, *Heimberger v. City of Fairfield* concerned a man successfully rescued from his burning home, where he had been trapped. He sued for damages, alleging that he suffered rough treatment while being saved. In that case, the court applied the immunity for negligence in fire fighting to rescue efforts while fighting a fire. This extension of immunity was limited to rescues concurrent with instances of fire; this same court ruled that rescue efforts independent of fires are not subject to statutory immunity. See the discussion of statutory immunity involving rescues, specifically the case of *Lewis v. Mendocino Fire Protection District*, in Chapter 6.

Acting From Private Motives

When a fire fighter exercises power for a nonpublic purpose, to someone's detriment, the fire fighter is liable for any resultant injuries. For example, the motive in sounding the siren and preempting the right of way might not be to respond quickly to a fire, but to speed a friend to the airport. Immunities and other affirmative defenses generally will not apply under these circumstances.

Acting Recklessly

In general, when a fire fighter uses his power willfully or recklessly, to the detriment of others, he will not be protected by immunity statutes. Suppose, for example, that without the existence of an emergency call, a fire fighter activates audible and visual emergency warning devices that permit him to disregard many rules of the road. If he then runs down a vehicle which had failed to heed his warnings, he cannot claim immunity from liability. As most immunity statutes provide protection only for acts of negligence, protection does not extend to acts of gross negligence or willful misconduct.

An example of how this principle has been written into state law can be found in California Government Code Section 850.8 relating to transporting people injured at a fire. The section states that a fire fighter or other public employee, when acting in the scope of his employment and when the injured person does not object, may transport or arrange for the transportation to a physician or hospital of any person injured by a fire. Further, the employee will not be held liable for any injury sustained by the injured person as a result of such transportation, or for any medical, ambulance, or hospital bills incurred by or in behalf of the injured. In addition, the employee shall not be liable for any other damages, unless they were caused by his willful misconduct in transporting the injured person or arranging for such transportation.

In summary, whenever a fire fighter departs from the terms of the power granted to him as a public officer or employee, he becomes responsible for his acts just as if he had never been authorized to act in a public capacity.

Discretionary Powers and Ministerial Duties

Another matter which deserves consideration is the distinction between nonliability for failure to exercise, or for negligence in exercising, a discretionary power, and liability for failure to perform a ministerial duty. Problems occur in determining what act or activity constitutes the exercise of discretion. "Many officers or employees carrying out the functions entrusted to them by others must frequently assess facts and choose how to act or not to act upon them. But not every exercise of judgment and choice is the exercise of discretion. It depends on the kind of judgments for which responsibility has been delegated to the particular officer. Discretion, as this court has noted, involves room for policy judgments." (*Dalehite v. United States*)

"An officer or employee is not engaged in a discretionary function or duty whenever he or she must evaluate and act upon a factual judgment. Discretion . . . exists only insofar as an officer has been delegated responsibility for value judgments and policy choices among competing goals and priorities." (*Nearing v. Weaver*) An example is where the fire chief or his representative has the power to perform fire inspections, but the time, place, and manner of inspection are chosen by the officer. Then he typically is performing a discretionary activity, with attendant immunity from liability for the exercise, nonexercise, or negligent exercise of the power.

Discretionary activities are distinct from ministerial activities, which characteristically do not involve the exercise of discretion. Ministerial duties are certain, fixed, and proscribed by rule, ordinance, or regulation. Their performance is required. The officer has nothing more to do than determine the facts which give rise to his duty to act, so no discretion is involved. When the facts have been ascertained, the individual is required to act—for instance, to grant or deny a license, to undertake specified proceedings, or to notify some authority. If he refuses to act one way or another, he would be subject to a judicial *writ of mandamus*, an order to compel performance under penalty of contempt and exposure to civil liability for any damages that result.

Therefore, an officer is liable for nonperformance or negligent performance of a duty that is clearly set forth, where the means and ability to perform the duty are present and within his control. As stated in *Doeg v. Cook*: "In short, where the duty is plain and certain, if it be negligently performed or not performed at all, the officer is liable at the suit of a private individual especially injured thereby."

Liability of Superior Officers

Under the doctrine of *respondeat superior*, the "master" is liable for the torts committed by his agent or servant while acting within the scope of his employment, even if the servant acts contrary to the master's instructions. Where the master participates in the negligent act or directs his servant to perform it, the master is liable. Such liability extends to the malicious acts of the servant, as well as to the negligent ones, unless they are committed exclusively for the employee's own purposes. In the absence of *respondeat superior*, and without strict liability, there is no charge of liability of the superior—the master—for the acts of the servant.

Maintenance of a discriminatory work environment can subject superior officers to personal liability in addition to the liability of the fire department. In *Hamilton v. Rodgers*, a discharged radio technician of the Houston Fire Department brought suit against the city and his supervisors for racial harassment and retaliation. In his complaint, he alleged that he was the subject of "ugly racial slurs and nasty pranks" and that, as a result, his health materially deteriorated. He claimed that, despite his favorable performance record, the hostile environment and resultant health change caused his suspension, although he later was reinstated in a lower paying position. In holding the supervisors liable, the trial court noted that not only did the supervisors ignore the racist antics of the plaintiff's co-workers but themselves discriminated against him. "The [two] supervisors tried to freeze out plaintiff by refusing to provide him with the assistance that he needed to become proficient at radio maintenance.... One may plausibly conclude, moreover, that discriminatory intent motivated decisions such as those denying plaintiff a car assignment and scheduling him for night shift." The court thus deemed it appropriate to hold the supervisors liable.

Need for Insurance

Fire fighters who drive public vehicles should take out personal insurance. The insurance can be either a separate broad form policy or a rider on another policy to protect themselves in case they injure someone by driving while acting outside the scope of their employment.

For example, a city vehicle is driven in a supermarket parking lot by a fire fighter shopping for a personal item. While parking, he negligently damages another car. Under the circumstances the city attorney is likely to find that the fire fighter's activity was outside the scope of his employment. The attorney therefore will refuse to defend the fire fighter in a lawsuit or pay any resultant judgment. He may even require the fire fighter to pay for damages to the city vehicle and may prefer disciplinary charges against the fire fighter for unauthorized use of a government vehicle.

Many insurance policies are not in effect while the insured is driving a government vehicle. In addition, if a fire fighter regularly drives his own vehicle on city business, he may not be covered by his own automobile liability policy. For example, a fire fighter leaving the engine house and using his own car or a department vehicle to go to a store to buy something for the fire station would be acting within the line of duty and be covered by insurance. However, if he leaves and shops exclusively for items for personal use, whether driving his own or a fire department vehicle, he is acting outside the scope of his authority. For either type of situation, he should be protected by insurance.

Sample Problems

Having reviewed the rules of liability that are applicable to a fire department, its supervisors, and fire fighters, let us look at a specific example of a common problem: the possibility of liability for permitting civilians to ride in fire department vehicles or for allowing them to engage in athletic activities at a fire station.

From time to time, most fire departments receive requests from people outside the department seeking permission to play on department-maintained recreational facilities, to temporarily reside in a fire station, to ride to fires on apparatus, or to participate in training programs. It always is good practice to require such guests to sign a "visitor's waiver" (see sample copy in the Appendix to this chapter) in which they agree to relinquish all claims of liability against the city, the fire department, or any individual member for personal injuries sustained while engaged in such activities.

Such waivers have questionable validity, because advance waiver of all liability does not bar anyone from filing suit if he is injured through the negligence of someone else. It should be noted that signed waivers can help in a municipality's defense to show that the plaintiff was aware of the hazards and risks associated with his activities. For this reason, it is desirable to require persons who are not members of the fire department to sign visitor's waivers prior to engaging in any department activity involving any risk of injury.

Insofar as liability of the fire fighter is concerned, it should be remembered that anyone can be held liable for injuries sustained by another person:

1. where he deliberately caused the injuries to happen, or
2. where he negligently caused the injuries, and the injured person was not contributorily negligent. Note, that states that have adopted the

comparative negligence rule hold that contributory negligence is not a bar to a favorable judgment; rather it only serves to reduce the allowable recovery.

Therefore, a fire officer or fire fighter who permits a visitor to play handball at the station generally will be held not liable to the visitor unless he deliberately does something which causes an injury or is actively negligent toward the visitor.

Strikes

Strikes—What Are They?

A "strike" has been defined as "the cessation of work by workmen as a means of enforcing compliance with a demand made on their employer." This definition implies a complete severance of relations between the employees and the employer. (In re *Webb*) However, the meaning of "strike" is not limited to that definition. Actions that fall short of totally stopping management-employee relations can be considered strikes. In the case of *City of Dover v. IAFF*, fire fighters stated that they would not return (call back) for any bell alarms or mutual aid calls until the "pay situation has been resolved." They were considered to be undertaking a strike action. This was because, despite their off-duty status, the regular and universally understood custom necessary for adequate community fire protection was not followed. Custom called for fire fighters to voluntarily return for general alarms, so their concerted failure to respond was considered a strike action.

Is There a Right to Strike?

There is no uniform national rule pertaining to the right of public employees to strike. However, the general rule, both by common law and by statute, is that public employees do not have a right to strike. "The primary policy implicit in the Public Sector Labor Law is to ensure the uninterrupted provision of services to the public vital to its health, safety, morals, and welfare. To that end, the strike is withheld from public employees [by common law]. To that end, the exercise of the strike by public employees has been declared illegal under . . . statute, and to that end other protest by public employees which impairs the public health or safety is enjoinable." (*State of Missouri v. Kansas City Fire Fighters*) Thus, in most states and under the Federal system, public employees do not have the right to undertake a collective work stoppage action against a public employer.[6]

---

[6] See "Damage Liability of Public Employee Unions for Illegal Strikes," 23 *Boston College Law Review* 1087 (1982)

On the other hand, a small minority of jurisdictions hold that public employees do have a right to strike. However, the right is a limited one: where the public employer can demonstrate that continuation of the work stoppage is directly and materially detrimental to public health and safety, the strike action may be restrained by order of court. Courts in such cases hold that public employees, having exercised their constitutional right to quit work, are no longer under their employer's control. Thus the employees cannot be ordered to perform duties incident to their former employment and cannot be found in contempt (and jailed) for refusing to work.

Note that the individual union member has a right not to strike or a right to resign from the union during a strike. Such actions cannot serve as a basis for union discipline. For example, in *Pattern Makers League v. NLRB*, the constitution of a national union prohibited resignations during a strike. When an employee was fined for resigning, the union's activities were held to be an unfair labor practice—an unlawful act of coercion against the employee's right, under the National Labor Relations Act, not to engage in a strike.

What Is the Liability for Striking?

Consideration of the question of liability for illegal strikes must distinguish between a suit instituted by a public entity and a suit brought by a private plaintiff. Where the plaintiff is a private citizen or entity, the majority of legal decisions hold that there is no right of recovery against the collective bargaining unit (such as an employees' union or association) for illegal strike activities. As stated in *Fulenwider v. Fire Fighters Association*, an injured property owner is considered only an incidental beneficiary of the normal activities of fire fighters. Benefit of fire fighter services is directed at the community as a whole, and not the individual property owner. For this reason, there is no cause for action by the property owner due to failure of the fire fighter collective bargaining unit to comply with law or the collective bargaining agreement.

When the plaintiff is a public entity, a majority of states permit recovery of compensatory and punitive damages against the union and its leadership for the costs of maintaining fire protection services during the period of a strike. For example, many states permit public entities to recover damages against the union under authority of statute.[7] Various other states have permitted recovery based upon theories of tortious interference with contract, public nuisance, *prima facie* tort, and a *per se* tort.

---

[7] See Minnesota Statutes Annotated Section 179.68.1 (1983), New Hampshire Statutes Annotated Section 273-A:15 (1977), Wisconsin Statutes Annotated Section 111.89(2) (c) (1974), Florida Statutes Annotated Section 447.505 (1981), Iowa Code Annotated Section 20.17(5) (1978), and Kentucky Revised Statutes Section 345.100(3) (1977).

In *State of Missouri v. Kansas City Fire Fighters*, the court sustained the right of the state to recover costs of activating the National Guard for the period of the illegal strike. In this case, the union obtained agreements from fire fighters of neighboring communities that they would not respond. Then the union went out on strike despite a court restraining order against the activity. The National Guard was ordered into the city to provide protection, resulting in an expenditure of $128,782.72 for Guard operations. After failing to obtain a satisfactory judgment based upon breach of express or implied contract, an action for recovery was instituted based upon the theory of tortious violation of law. Under the latter theory, the court sustained the state's right to a compensatory and punitive damage recovery against the union, its officers, and directors. However, recovery against individual members of the union was prohibited. This was because there was no evidence that each individual member affirmed or ratified the action of the majority of union members or the union's Board of Directors. "Liability does not devolve upon a member from the fact of association only, but from a personal act, or from the act of an agent whose agency must be proven. Thus, an association member sustains no personal liability for the acts of the officers or other members unless the member participates in, authorizes or ratifies the acts," said the court in this case.

A fire chief who directly interferes with activities of fire departments assisting during a strike within his own department could be personally liable for any resultant damages. For instance, property owners sued the union and city's fire chief for threatening mutual aid fire fighters and blocking their entry into a building on fire. When the chief's action resulted in total destruction of the building, the court held that the chief could be held personally liable. (*Berger v. City of University City*)

Picketing—Is It Legal?

Picketing is considered a valid method of expression, so ordinarily an injunction will not be granted against the peaceful picketing of fire stations or fire scenes. However, if it could be shown that the picketing of the fire station or fire scene interferes with the city's continued fire suppression duties, picketing could be prohibited. When failure to disperse results in damages incurred by the city or a third party, the situation could give rise to a "for cause" action for damages.

Review Activities

1. What is the duty of the EMS dispatcher? What would your recommendation be if the need for an ambulance were questionable? What are your suggestions for department guidelines on the subject?

2. In your own words, define the "duty of reasonable care" of a fire fighter, an EMT, and a fire officer. Against what standards is each duty measured? Explain.
3. Describe the rules that govern participation of civilians in recreational activities at your department. What liability exposures do you consider to be involved? What would be your recommendations, if any, for change? Why?
4. State the circumstances under which the following individuals would have a liability exposure during a strike:
   A. Union President
   B. Fire Chief
   C. Union Executive Board Members
   D. City Manager
   E. Fire Fighters
5. Define the "Good Samaritan" law. Does your state have one? What is the effect of such a law?

Appendix to Chapter Thirteen

VISITOR'S WAIVER

To: Fire Chief, ____________ Fire Department
From: ________________________

Request is hereby made for permission to: __________ (participate in recreational activities at fire department facilities; ride fire department apparatus; engage in fire department activities, etc.). Purpose of such undertaking being: ____________ (research, participation in department physical fitness activities, etc.). Requested duration of such permission being: ______________.

In consideration of your granting such permission, I agree to and do hereby waive any and all claims against the City of __________, its Fire Department, and all the members and employees of said department for any injury or accident occurring to me while on the fire department facilities or equipment, or at fire scenes. This waiver shall be and is binding upon me, my heirs, assigns, and personal representatives.

Signed: ______________________
Date: ______________________

Permission is hereby (granted/denied) for ____________ (name of applicant) to participate in ____________________ (activity) for the period: ____________ through ____________.

Signed: ______________________
Title: ______________________
Date: ______________________

Distribution: Visitors File
Officer in Charge of Activity
Applicant

Chapter Fourteen

Procedural Pointers

Surveys in the Field

The success of a fire prevention survey or inspection often depends upon the manner in which it is conducted, the level of communication employed, and the way any recommendations are presented to the owner. Where fire prevention is perceived as a mutual concern of the fire department and the owner, for the safety of employees and the public, major voluntary strides toward achieving safety improvements can result. However, where the concern is simply to see that the owner complies with the "code," without any explanation of why the improvement is required—why the code requires the action—hostile owner reaction may result.

Fire inspectors find that people may resent the idea that they must be told, in writing, to do some simple thing which they readily agree to do right away. If the situation can be corrected while the inspector is still on the premises, it is good practice to encourage such prompt action. However, if the violation requires time to remedy because of its complexity or magnitude, the inspector should prepare a statement showing just what the hazardous conditions are, and what corrective action is required. References to the building or fire code provisions that cover the issue should be included. Supplemental drawings, photographs, or sketches which describe the conditions as they are or as they should be also may be given to the owner. The statement should not tell the owner how to perform the work, only what the final result should be. The statement should be in a form readily understood by an architect, builder, or fire suppression system contractor, and in enough detail so another inspector, unfamiliar with the conditions, could effectively respond to inquiries if necessary.

Written Notices

Where conditions do not warrant an elaborate survey but are hazardous enough to necessitate development of an inspection record, fire inspectors might write a "notice" to the owner or occupant. Such a notice should concisely outline the fire protection standards whose requirements should be met, with exact code references. Sometimes such a notice is printed in the form of a "recommendation memorandum," thus avoiding the impression of giving a "ticket" or ultimatum on the first inspection visit. If, on reinspection, it is found that there has been no compliance without justification, it is customary to give the occupant or owner a "final notice." If this is not obeyed within the stipulated time period and the hazard is not serious enough to require on-the-spot corrective action, department procedure may call for giving a "formal notice" to the violator. A formal notice would direct him to appear in the office of a deputy city attorney or district attorney, at a specified date and time, when he will be asked if he has any good reason why a complaint should not be filed against him. Often this procedure and hearing are sufficient to clear up any misunderstanding and achieve code compliance.

Thomas Williams, a former Deputy City Attorney of Los Angeles, has described some of the fundamentals involved in obtaining enforcement of fire safety ordinances. The fundamentals summarized in succeeding paragraphs are based upon Mr. Williams' descriptions, augmented by the author's own experiences.

The Inspector's Demeanor

The inspector's demeanor has an important bearing on the potential success of fire prevention efforts. He should realize that his inspection could be the first direct contact of the municipality with a property owner or manager. While the inspector's uniform and bearing should command respect, he should not overlook the adage that "the greater the man, the greater the courtesy." Through a combination of tact, a thorough working understanding of fire prevention principles, appropriate consideration of the other person's problems, and his own willingness to determine all the facts, the inspector will gain the respect of the public as well as of his peers. He should be objective, never showing agitation over matters. In writing his notices, he should clearly explain what he wants done. Each notice should be complete, listing all the material deficiencies on the original inspection memorandum—not just one or two items. Avoid subsequently listing additional items when enforcement by citation or legal process is required. It is embarrassing to have to explain why the additional items were "overlooked" in the original memorandum.

The inspector should be careful to give the notice to the person responsible for the violation—the owner, occupant, or some third party. Because he may be called upon to identify the defendant in court, he should be sure to determine the identity of the violator when giving notice.

In cases where the violator refuses to carry out the inspector's recommendations, and where the interview with the municipal attorney has not produced the desired results, the inspector ordinarily requests that a complaint be issued. Sometimes the municipal attorney will have the complaint issued immediately following the hearing, if the results are not fruitful. The inspector should be prepared for this possibility and offer his recommendation, pro or con, concerning such action. Note that a complaint cannot be issued automatically. Rather, the attorney involved must satisfy himself of the existence of *prima facia* elements of the offense—evidence reasonably sufficient to find that the offense occurred and was committed by the defendant. By law, the attorney must verify that all elements of the offense are present, to avoid a charge of malicious prosecution. The inspector should be fully prepared to assist the attorney in verifying these requirements. By providing such assistance, he will go a long way toward developing the inspector-attorney rapport necessary for establishing an effective code enforcement program.

Authority to Make Arrests

Regarding the authority of fire fighters to make arrests, bring in witnesses, make necessary investigations, and file complaints, several points should be noted. A member of the fire department has the same rights that any citizen has to make an arrest. He may do so: 1. whenever a public offense is committed or attempted in his presence; 2. when he finds that a felony—arson, for example—has been committed and he has reasonable grounds for believing that the person arrested has committed the offense; or 3. when he knows that the person he is about to arrest has committed a felony, although not in his presence.

In many states, the circumstances under which, and the time when, an arrest may be made are further prescribed by statutes. For example, a statute may limit performance of an arrest for a misdemeanor offense to between the hours of 6:00 a.m. and 10:00 p.m., unless performed in a public place. The law also can prohibit performance of a misdemeanor arrest where it requires forcible entry into the residence of the defendant, such arrests generally being reserved for peace officers. Where there is any question as to authority to perform an arrest, it is best to inform a peace officer of the circumstances and leave any arrest action to him.

How to Make an Arrest

To make an arrest is to take an individual into custody in the manner prescribed by law. An arrest is accomplished by placing the person under restraint or control, or by his submission to the custody of a peace officer. The individual need not be physically restrained, for custody will have occurred when the individual subjectively believes that his freedom of movement has been removed. Under no circumstances should restraining force be greater than that reasonably necessary to take and keep control over the individual.

In carrying out the arrest, the person must be informed of the intention to arrest him, the cause of the arrest, and the authority to make the arrest. An exception is when the person to be arrested is engaged in the act constituting the offense or is being pursued immediately following commission of the offense. If the arrest is being made under a warrant, the warrant must be presented and the person being arrested must be provided with an opportunity to read it.

In the case of a felony arrest, should the person flee or forcibly resist, the arresting officer may use all reasonable force to carry out the arrest. However, deadly force may not be employed unless the person attempting escape poses a significant threat of death or serious physical injury to the arresting officer or others. (*Tennessee v. Garner*)[1] It is not proper to shoot a person who is fleeing from arrest. For example, say the offense committed is only a misdemeanor. The arresting officer shoots the fleeing person to obtain his submission. The officer in this case would be guilty of the criminal act of battery should he hit the fleeing person, assault if he misses.

Informing a Suspect of His Rights

To preserve the admissibility of evidence and statements obtained following an arrest, it is critical that the defendant be informed of his constitutional rights. As the United States Supreme Court stated: "The Fifth Amendment privilege [against self-incrimination] is so fundamental to our system of constitutional rule and the expedient of giving an adequate warning as to the availability of the privilege so simple, we will not pause to inquire in individual cases whether the defendant was aware of his rights without a warning being given. Assessments of the knowledge the defendant possessed, based on information as to his age, education, intelligence, or prior contact with authorities, can never be more than speculation; a warning is a clear-cut fact. More important, whatever the background of the person interrogated, a warning at the time of the interrogation is indispensable to overcome . . . [the arrest] pressures and to insure that the individual knows he is free to exercise the privilege at that point in time." (*Miranda v. Arizona*)

---

[1] This case says that the use of deadly force to prevent the escape of all felony suspects, whatever the circumstances, is constitutionally unreasonable. Only where a real threat of death or serious bodily harm exists is use of deadly force justified.

The court's inquiry will be limited to whether or not the warnings have been given and understood. The net effect of the rule is that the prosecution may not use any statements or evidence stemming from custodial interrogation unless it is demonstrated that the constitutional warnings were given prior to any interrogation. This restriction applies whether the evidence is exculpatory or inculpatory—tending to excuse or blame the person under arrest.

The Specific Warnings

"Prior to any questioning, the person must be warned that he has a right to remain silent, that any statement he does make may be used as evidence against him, and that he has a right to the presence of an attorney, either retained or appointed. The defendant may waive effectuation of these rights, provided the waiver is made voluntarily, knowingly and intelligently. If, however, he indicates in any manner and at any stage of the process that he wishes to consult with an attorney before speaking, there can be no questioning. Likewise, if the individual is alone and indicates in any manner that he does not wish to be interrogated, the police [or investigating fire officer] may not question him. The mere fact that he may have answered some questions or volunteered some statements on his own does not deprive him of the right to refrain from answering any further inquiries until he has consulted with an attorney . . . ." (*Miranda v. Arizona* and *Edwards V. Arizona*)[2]

Admonitions

In addition to the specific statements of rights and waivers quoted above, the following admonitions are appropriate:

1. All statements—with very few exceptions—are subject to the Miranda warnings requirement;
2. If warnings are not given, statements and all resulting evidence—unless obtainable on fully independent evidentiary grounds—are inadmissible;
3. A defendant must be warned regardless of his knowledge of his rights. Where any material period of time has elapsed between the original warning and the initial interrogation, a second warning is appropriate;
4. His waiver must be knowingly and intelligently made. His physical and mental condition should be documented to show that he is giving a fully voluntary, knowing, and intelligent waiver of his rights;

[2] Police must stop all questioning at the first request for consultation with a lawyer. Also, once a defendant has invoked his constitutional right to counsel, further questioning without counsel can proceed only when the accused initiates further discussions with the police and has knowingly and intelligently waived the right he had earlier invoked.

5. It is appropriate to ask the defendant if he understands the effect of his waiver. Any affirmative response should be documented;
6. Questioning must stop whenever a subject indicates a desire to remain silent;
7. If a suspect wishes to talk without counsel, his statement—and the Miranda warnings—should be tape recorded, if possible. As an alternative, a hearing reporter should be present and the suspect should sign and date the transcription of his statement;
8. If a waiver is signed, the peace officer should evaluate the suspect's education, mental and physical condition, and any other factors to corroborate that it is a "knowing and intelligent waiver"; and
9. Lengthy interrogation or incommunicado incarceration before a statement is made is strong evidence that any waiver will be judged invalid.

Exceptions to the Need for Warnings

An ordinary traffic citation does not require a warning, as no "custodial interrogation" is being performed. In addition, no warnings need be given in the following circumstances:

1. When a person walks into the police station and states that he wishes to confess to a crime. However, if interrogation follows to detail the crime, warnings would be required;
2. When a person calls the police to offer a confession or any other statement; and
3. When an officer is engaged in general, on-the-scene questioning concerning a crime, as part of the fact-finding process. When, however, the inquiry is focused on a particular suspect and the purpose of the inquiry is to gather evidence that he was responsible for the crime, the Miranda warning must be given.

Interrogation/Search

If a fire investigator stops a person for questioning—for example, at the scene of a suspicious fire—and the person refuses to answer any questions, his action cannot be considered a breach of the peace that would permit a warrantless arrest. A person's exercise of his constitutional rights cannot serve as a basis for a criminal complaint. Nor can a person be arrested and searched without there being "probable cause" to believe that a crime has been committed and that the suspect was responsible. However, where a person is stopped based upon a "reasonable suspicion" of his potential involvement in a crime, and a "reasonable suspicion" is developed that the suspect is armed, a "pat down" search for weapons is permissible.

Where "probable cause" did not exist at the time of arrest but where subsequent search of the suspect provided "probable cause"—such as his possession of an illegal concealed weapon—the results of the arrest and search will be held inadmissible. The reason is rooted in the fundamental legal maxim: "The fruits of an illegal search cannot be used to justify a prior illegal act." Note that any person making an arrest may take from the person arrested, and from his immediate vicinity, any offensive weapons or evidence of a crime. These should be presented to the magistrate, along with the suspect, at time of arraignment.

Detention of Witness—False Arrest

A private individual who has arrested someone for the commission of a public offense must, without unnecessary delay, take the person before a magistrate or deliver him to a peace officer. If the person is not placed under arrest, but voluntarily consents to go to an office for an interview, this is proper as long as he is not compelled to submit to any questioning and is free to leave at any time. Except in highly special cases or investigatory activities of grand juries and other governmental inquiry bodies, there is nothing in the law that authorizes peace officers or anyone else to act without a warrant to "bring in a witness" under the legal process for the purpose of securing testimony or information. The forceful detention of a person not lawfully arrested constitutes a false arrest, for which peace officers as well as private citizens may be held accountable. Moreover, failure to promptly take the arrested person before a judge or magistrate is considered a criminal misdemeanor, separately punishable.

Warrants

The two principal types of warrants are arrest and search. Both must be issued by a neutral judge or magistrate, based upon a showing of "probable cause." It is the duty of the neutral judge or magistrate to determine the existence of probable cause based upon the facts presented. Its existence is based upon an examination of the "totality of the circumstances," giving significance to each relevant piece of information and balancing indications of the relative reliability and unreliability of the information. (*Illinois v. Gates*) There is no rigid formula focusing upon the intended purpose of the warrant.

Probable cause for an arrest warrant exists where there is adequate reason to believe a crime has been committed and that the suspect is the perpetrator—the one who committed the crime or had it done. An arrest warrant directs a peace officer to arrest the named or described individual and present him before a court to initially respond to the charges against him. Where the individual cannot be identified, a "John or Jane Doe" warrant is issued; it describes the person to be arrested with as much specificity as possible. Upon arrest, a limited search of the individual and his immediate vicinity, for offensive weapons or evidence of a crime, is permissible. If the warrant is valid on its face, the arresting officer is protected from a false arrest suit provided he carried out the arrest properly.

A search warrant authorizes a peace officer to search a designated premises, area, or item for evidence of a crime, illegal fruits of a crime, or contraband items. Probable cause for a search warrant exists where there is adequate and reliable evidence to justify a reasonable belief that evidence of a crime, or contraband, is present at a specified location. Search warrants do not specifically authorize the arrest of the person possessing the evidence; however, an arrest can be carried out whenever possession of the item itself constitutes a crime, such as possession of controlled substances. The authorized search is limited to the area and location described in the warrant. Illegal or evidentiary items noted "in plain view" during the search can be legally seized, however. Upon completion of the search, the executed warrant must be returned to the issuing magistrate. It should detail the items seized and their disposition.

Malicious Prosecution

Filing a complaint against a person, or beginning an action—whether civil or criminal—against him without probable cause (reasonable grounds) is considered "malicious prosecution." Where such action has occurred, the law permits recovery for all damages suffered by the defendant. Care should be taken to verify the factual basis that underlies a complaint, to make sure there is probable cause to proceed.

Arson

Some fire departments are charged with responsibility for investigating suspicious fires, interviewing witnesses, collecting evidence, arresting suspects, and assisting the prosecutor in the preparation of cases for trial. Several books cover the subject of arson, so the discussion here will be limited to a few basic observations. Relevant topics, such as search and seizure, interrogation of witnesses, waivers and warnings, and arrest procedures already have been discussed in this chapter.

At common law, arson was the felony of maliciously and willfully burning someone's dwelling. Because it was an offense against the security of a person's habitation, it was considered a crime of greater enormity than other acts of unlawful burning. Setting a house fire was considered an aggravated felony because such fires exposed individuals to personal injury and showed a greater recklessness and contempt of human life than the burning of any other building. The crime is now statutory in most jurisdictions. Although some penal codes define the crime substantially the same as it was at common law, some statutes have extended coverage of the crime to include unlawful burning of barns, hayfields, bridges, and even vehicles.[3]

In addition to state statutory arson laws, Federal prohibitions against intentional fire-setting are playing an increasing role. For example, under Title 18, United States Code, Section 844(i), it is a Federal crime to "maliciously damage or destroy, or attempt to damage or destroy, by means of fire or an explosive device, any building . . . used . . . in any activity affecting interstate or foreign commerce." The extensive scope of the Federal law includes any building or structure that in any way affects interstate commerce. In *Russell v. United States*, the U. S. Supreme Court affirmed the conviction of a property owner for burning his apartment house. The court held that Congress intended by the law to protect any and all business directly, or even remotely, engaged in interstate commerce; and the housing occupied by workers engaged in interstate commerce activities could be considered protected under the law.

Elements of the Offense

Two elements must be proven beyond a reasonable doubt in order to sustain a conviction for the crime of arson: 1. the *corpus delicti*—such as a fire intentionally and maliciously caused, counseled, or procured by a criminal agency; and 2. the identity of the defendant as the one responsible for the fire. The mere burning of a structure does not establish the corpus delicti, for there is a rebuttable presumption that a fire is of accidental origin. To constitute arson, the burning must be the result of the willful and criminal act of some person. Even though there may be *prima facie* evidence that a fire was of incendiary origin, such as the finding of separate and distinct fires on the premises (*People v. Sherman*), and even though only one person was known to have been on the premises at the time of the fires, the presumption of innocence accorded to that person would have to be rebutted in order to prove the second element of the crime of arson—identification of the arsonist.

Circumstantial Evidence

The very nature of the crime of arson involves secret preparation and commission seldom observed by eyewitnesses. This makes it difficult for the state to produce eyewitnesses to directly establish the guilt of the accused. Instead, many if not most cases establish proof of guilt by relying on circumstantial evidence. In a Utah case, *State v. Showaker*, a conviction was sustained where the defendant, when fired, had threatened to burn the business; where no trace was found in the building of office furniture associated with the defendant; and where finger and palm prints on an overturned drum of

---

[3] As a result of *State v. Baker* in 1903, the arson statute in Wisconsin was extended to include the intentional burning of an automobile. Since then, many arson statutes in the U.S. have been revised similarly.

acetone on the premises were identified as his.[4] In the Wisconsin case of *State v. Kitowski*, a verdict of guilty was returned by the jury against a man who had been seen leaving the house of a girl friend just before fire was discovered there. He was sighted shortly after he had announced, in a fit of anger, that he was going to burn her house to the ground. To sustain the defendant's conviction for arson, the "well settled standard of proof in circumstantial evidence cases [is] that facts from which the inference of defendant's guilt is drawn must be inconsistent with the defendant's innocence and must exclude to a moral certainty every other reasonable hypothesis." (*People v. Marin*)[5]

Intent and Motive

In a legal sense, "intent" is quite different from "motive." Intent may be defined as the design, resolve, or determination with which a person acts[6] to effect a certain result, while motive is the reason which leads the person to desire that result. (*Baker v. State*)

Ballentine[7] defines "criminal intent" as "that evil state of the actor's mind, accompanying an unlawful act, in the absence of which no crime is committed." As pointed out in Chapter 3, in the discussion of criminal prosecutions, in every crime there must exist a union, or joint operation, of act and intent. In a prosecution for arson, the criminal intent may be inferred from the defendant's acts or statements, or from circumstances connected with the offense.

In criminal law, according to Ballentine, motive is "that which leads or tempts the mind to indulge in a criminal act." Motive may be inferred not merely from the attendant circumstances of the offense, but from all previous occurrences which have reference to, or are connected with, commission of the offense.

Motive, then, is the cause or reason which induces a person to commit a crime. While its determination may be helpful in establishing the guilt of the defendant, it never is legally necessary to prove motive in order to obtain a conviction. Arson investigators must look for motive, however, for it is useful in helping to explain the actions of someone charged with arson, and even in connecting the defendant with the offense for which he is charged.

What is "Burning"?

To have the crime of arson, there must be a "burning" of something. The crime is complete in this respect if the nature of the material fired is changed

---

[4] See *Arson Reporter*, Vol. 7, No. 9, Sept. 1986, and other issues for further

[5] Defendant's conviction on 26 counts of murder in the second degree and arson in the fourth degree was reversed for failure to fully prove the arsonist's identity.

[6] Black, Henry C., *Black's Law Dictionary*, 5th Edition, West Pub. Co., 1979

[7] Ballentine, James A., *Law Dictionary with Pronunciations*, Lawyers Co-operative Pub. Co., 1930

or charred. A wasting of the fibers of the wood, no matter how small in extent, is sufficient to prove arson if the other elements of the crime can be established. "The law is well established that a charring of the fibers of a part of a building is all that is required to constitute a burning sufficient to make the crime of arson. Any charring is sufficient." In *State v. Nielson*, the question was raised of whether or not an acoustical tile could be considered part of a building. When burned, would the tile constitute the arson crime element of "burning a building"? The court said: "We cannot see any reason to make a distinction between a wood panel ceiling and an acoustical tile ceiling. Each is an integral part of the building."

Polygraph Evidence

The amount of care which should be exercised in obtaining confessions, to make them admissible into evidence, already has been discussed. On the question of the admissibility of polygraph tests in criminal trials, the courts generally have ruled against them, holding that materials obtained during all stages of the polygraph examination cannot be used because of the difficulties in assuring their reliability and accuracy. These materials might include the pre-test interview, the examination itself, the post-test interview, and the test analysis. Also, juries might give undue deference to the results of a polygraph test, beyond that which is reasonably justified. (*State v. Grier*)

Procedures regarding the use and admissibility of polygraph tests are undergoing continuous evolution. At the present time, use of polygraph tests is limited to stipulated cases—those such as *Bragg v. State* where both parties granted permission for their usage as evidence. The reason was aptly put by a California Superior Court judge who stated that the reliability of polygraphs is almost totally dependent upon the character and experience of the operator. It is a science, he said, but one much like psychology, in which the "expertness" of the expert sometimes is open to question.[8]

Handling "Sit-ins"

Because policemen sometimes are reluctant to clear the exitways of restaurants and similar places of public assembly, for fear of being accused of "police brutality," fire fighters may be called upon to handle these situations. Should this occur, it is a good practice to request the dispatch of a fire prevention officer and a photographer to verify and document hazardous conditions. After verification, if a warning of possible arrest results in the clearing of exitways, no further action need be taken. If the warning is ineffective, then a head count should be made of everyone in the occupancy. This is the

[8] Adler, Marlene, "Polygraph Evidence Not Admissible," *League News*, Aug. 1972, p.3

way to determine if the permissible occupant loads of the structure and of available exitways are exceeded. Should this be the case, the *prima facie* elements of the offense have been found.

The fire officer should then announce two or three times, using a voice amplifier, that the offenders are creating a potentially dangerous condition for themselves and others, as well as breaking the law. The officer should give the number of the applicable ordinance. He should warn people that they will be arrested if they do not move. If they do not, he should announce that they are under arrest and should ask the police to carry them off to jail. The police can use reasonable force in restraining those arrested; if they resist arrest, the resisting action becomes a second chargeable offense, a felony. Remember that the blocked exits for which the arrests are made must be required exits. Be sure to obtain the names of witnesses who will testify that the announcements were made, and how. The photographs will be useful in court to establish the violation.

Carrying a Badge

The mere fact that a fire fighter carries a badge does not make him a peace officer under the penal code definition. Nor does an ordinance giving him the powers of a "police" officer necessarily bring him within the state law's definition of that position. The state law may specifically set forth certain classes of enforcement personnel, omitting any reference to fire fighters.

Failure to carry a badge has relatively limited effect upon a fire fighter's powers or activities. However, where presentation of identification is requested and is reasonably necessary to protect the public interest, failure to show a badge might prevent enforcement of the law in question. The public's refusal to admit a fire fighter who carried no ID most likely would be upheld as a reasonable exercise of a person's personal safety interests. An example would be a building occupant's refusal to admit a fire fighter requesting entry for inspection purposes, the refusal being made on the basis that he could not show any identification.

Signing the Complaint

A criminal complaint is signed by the person making the complaint in his capacity as an individual, and not as an official. Legally, it is immaterial whether a complaint is sworn to by a police officer or a private citizen. This means that members of the fire department stand on the same footing as do members of the police department with respect to formally filing a criminal charge.

Hearings Before the City Attorney

In hearings before the city attorney, the fire inspector must be impersonal; he should state the facts and avoid giving the impression that the prosecution involves a personal issue. The inspector should not argue with the citizen who has been ordered to appear at the hearing.

Trial Preparation

Before a city attorney acts on a complaint, he often will request a statement of facts from the fire inspector. Such a statement should be as complete as possible, containing the same information that was presented to the accused in the written notices and final notices already issued. The statement might contain such facts as the building condition and specific violations. A verified statement, showing that conditions have not changed and explaining the importance of correcting the conditions for fire protection and life safety purposes, also may be requested by the city attorney.

The inspector must recognize that trying code violation cases is not considered a very challenging aspect of a prosecutor's career, so such cases frequently are given only limited attention. Also, because most prosecutors never see the file on a case until it comes to trial, the care with which the statement is developed will be critical to a successful outcome. This is especially true where the statement can quickly and completely give the attorney the facts and show the importance of obtaining a guilty verdict. The statement should stand alone and provide the attorney with enough information to fully support his case. All the facts must be stated, along with notations on all conversations with the violator and copies of all correspondence. All the documents should be in a neat, easily retrievable, and well-organized form.

If the violator cursed the inspector, the exact words should be used in the statement. Then if the violator pleads guilty or is convicted, the judge can note the statement as part of his determination of an appropriate fine or other penalty. Note that before quoting profanity or giving other testimony which may be embarrassing to jurors, the judge and jury should be alerted. Because of the potential prejudicial effect of an overly profane statement, a judge might exclude the testimony. If he is alerted to the statement before it is made, he can immediately rule on its admissibility. This will forestall any subsequent claim of a mistrial because of undue prejudice.

Along with the written preparation for trial, the fire inspector should prepare his personal appearance and manner of presentation. As the court considers the inspector an "expert" in fire safety matters, he should present himself this way. A timely appearance in appropriate uniform, without excessive trappings of office such as ribbons, braid, and badges, goes a long way toward conveying the image of expertise—even prior to his testimonial presentation. Chewing gum, smoking, and letting papers bulge from pockets detract from the professional image, and should not take place within the courtroom.

In the Courtroom

The inspector who has been working on the case should act as an advisor to the prosecutor in the courtroom. The inspector can assist not only by outlining the building or fire code ordinances involved, but also by explaining the reasons for the requirements and the history of their promulgation. Also, the inspector can serve as a valuable reference source for explaining technical terms, such as "flash point," in language that can readily be understood by the judge and jury.

On the Witness Stand

The time waiting to be called to the stand is a perfect opportunity for the inspector to review his notes, concentrate upon the legal case, and be sure his emotions are under control. He should enter the courtroom in a purposeful and serious manner, intent upon performing his official duties. He should take the oath in a direct, forthright manner, speaking in a clear, confident voice while facing the court clerk. Although the witness chair frequently is referred to as the "hot seat," the inspector should sit fully erect, feet flat on the floor, with hands comfortably placed. While on the stand, avoid playing with personal objects like a ring, watch, or pen.

Again, maintaining the image of the expert by direct and candid responses to questions is important. An overly hesitant delivery of testimony infers a lack of factual or professional confidence. A glib, too-ready or talkative response projects the image of a person who knows all the answers to even the most difficult or technical questions. Juries tend to distrust both the hesitant and the over-confident types. Rather, the inspector should make clearly articulated, on-point responses to all inquiries. Where he does not know an answer, a frank "I don't know" or "I don't recall" is appropriate. Refer to an inspection report or code document if the answer is contained there. As a witness, the inspector should present testimony only in response to questions, when and as posed. He should not make additional remarks unless they are necessary to clarify his testimony. Finally, the inspector's presentation should exemplify personal disinterest in the identity of the defendant but professional interest in the courtroom results.[9]

"Horse-Shedding" the Witness

The above expression dates back to the days when the lawyer took his client out to the horse shed, behind the courthouse or bar, just before the trial. Together, they went over the questions the lawyer was going to ask his client on the stand. This would give the lawyer a last-minute opportunity to verify the responses the client would give and to see if there were any proposed changes in the client's testimony.

It is perfectly proper for fire department members to review with the attorney the testimony that will be given at trial. This does not mean that any-

---

[9] See Tilton, Dennis, "Taking the Stand," Film Communicators, 1981, for further tips on courtroom preparation and presentation.

one should color the story or change his viewpoint in response to attorney inquiries. Rather, the review opportunity can be used to determine the basic facts or opinions that underly the viewpoint, and to see how strongly they are held.

The person testifying should be told that if the defense lawyer asks whether or not he "discussed this case with anyone," the answer should be "yes." If he blurts out "Oh no, sir," thinking that is the appropriate answer, he undoubtedly will be confronted with the lawyer's inquiry: "Have you discussed this case with the Fire Chief or City Attorney?" He then would have to respond "yes," thereby getting caught giving contradictory statements—a position that could needlessly rattle the confidence of the inspector and diminish his stature in front of the jury.

Hearsay Evidence

The fire department witness should never volunteer information; this is a very dangerous thing to do. Items inadvertently referred to can often prejudice a case. Nor should the witness make speeches. However, if a question calls for a "yes" or "no" answer, he has the right to request the court's permission to explain the answer. He should not give any hearsay evidence—that is, testimony referring to statements made by someone outside the courtroom and not under oath. Such statements generally are inadmissible based upon the accused's constitutional right to cross-examine all witnesses testifying against him. Generally, only facts derived from an individual's own knowledge or observations are admissible in evidence.

The "Know-It-All"

No witness should attempt to give the impression that he knows all the answers. The jury is likely to distrust a person who is too glib and too ready with a positive answer to difficult or technical questions. The attorney for the defendant might ask the fire department investigator, for example, "What is the temperature at which nitrocellulose commences exothermic decomposition?" or "How many minutes will a 'king size' cigarette burn on top of a newspaper before being completely burned to ash?" If he doesn't know the answer, the investigator should say so frankly, or say he doesn't remember. Unless he is called as an expert witness (and, as such, should be prepared), the investigator should not even venture a guess as to the answer to such a question.

Avoid Specialized Expressions

Avoid the use of specialized vocabulary or "canned" expressions that sound like a reiteration of lecture notes from the training academy. Jargon of the fire service or specialized expressions in the fire code should be presented in plain English. Statements such as "Type V-N buildings cannot contain H-2 occupancies on floors below grade" only serve to confuse a lay jury.

Rather, make a clear statement like "Occupancies that use hazardous materials are not permitted in basements of non-fire-rated wood frame structures." Try to fully apprise the jury of the facts, in a form readily understood. Only where fire department jargon is material to the case should it be used in testimony, and then only with an explanation or "translation" into ordinary language.

Silence Is Not Always "Golden"

There are times also when a witness's silence may have implications—wrong implications. Silence can cause a judge or jury to draw the wrong inferences. For example, at a coroner's inquest following a fatal hotel fire, the fire chief might be asked the date of the last fire prevention inspection at the hotel. To give no answer may imply that the location had been overlooked—that no inspections had been made there. This inference would materially diminish the professional stature of the testifying chief. Clearly, important facts should be readily available, in either testimonial or record form.

Exhibits and Notes

It is good practice to bring in physical evidence and exhibits, to use a blackboard to draw diagrams, and to recreate physical action in an effort to prove the case. Photographs introduced with the proper foundation also are excellent aids. Notes taken at the scene or right after an incident should be brought to the courtroom, for use by the witness to refresh his memory, with the court's permission. The witness does not read his notes aloud, but after privately reviewing them on the stand and refreshing his memory, he proceeds to testify without them. In civil cases, the notes also can be admitted into evidence as a "past recollection recorded," even though the event occurred so long ago that the notes do not refresh the witness's memory. Because he made the notes at a time when the facts were fresh, he can testify as to their accuracy.

Impeaching a Witness

Another reason that keeping accurate notes is important is that they can be used to impeach or refute the testimony of another witness. For example, when a witness for the defense tells a story different from that of the inspector, accurate and complete notes could be used to refute the defense testimony. In another instance, the prosecutor could, unknowingly, impeach his own witness by having the inspector's testimony contradict a prior witness's account of an incident. This raises the question of who is telling the truth. Accurate notes can help avoid embarrassing the inspector. Note also that a witness may be required to present his notes to the jury; if they are at all inconsistent with the inspector's testimony, the effect on the jury would be highly detrimental. Cautious review of notes is advised.

Relevancy

In testifying, do not make statements which are not relevant to the case. For example, if the violator says to the inspector, "Do not file a complaint against me because I'm on parole and will go back to the pen," do not include such a statement in direct testimony. If such an overly prejudicial statement were uttered, the court most likely would declare a mistrial. While it is proper to testify that a conversation was held with the defendant, confine the testimony to what was said on matters relevant to the case—truly bearing on the issues involved. If on cross-examination the question is asked whether or not any other conversations were held with the defendant, then it is proper to say "yes" (if such were the case). If defense counsel asks what the defendant's exact words were, then it is appropriate to relate them, for the defendant's attorney has elicited the prejudicial testimony by his probing.

Do Not Stress the Immaterial

In addition to avoiding testimony containing facts not relevant to the case, do not give undue emphasis to matters which are immaterial. For example, it may be relevant that a defendant was storing high-octane gasoline in his warehouse. However, when the charge is unlawful storage of flammable liquids, it is not really material whether the liquid was gasoline or benzene, less material whether it was high- or low-octane, and almost totally immaterial whether it was manufactured by X or Y. Overly detailed descriptions may detract from the more important information the presecution wishes to present. The main point to be made is that the liquid had a flash point which brought it within the code definition of "flammable"—a fact that easily could be established by independent laboratory analysis.

"Sticks and Stones"

The witness should never be argumentative in his testimony, nor should he allow himself to get mad at the defense counsel. That is exactly what the opposing attorney would like, for argumentative testimony destroys the inspector's objective credibility. Rather, it is best to take plenty of time and to answer each question slowly and deliberately; the effect is likely to needle the attorney more than the witness. Should the inspector get truly agitated, it is best to ask leave of the court for a recess, possibly to visit the rest room, to provide an opportunity for the inspector to compose himself.

The Verdict

Once the decision is handed down, it is not wise to "take it to heart" if the case is lost, nor to badger the defendant with "I told you so" if the case is won. It is not fitting for a public official to be viewed as a poor loser; likewise, it is unprofessional to carry on as the triumphant winner. Rather, where the case is lost, it should be reviewed for lessons learned, ways it could have been improved, where further homework could have payed off, etc. Where it is won, it should be viewed as a fortunate experience that demonstrates the benefit of thorough preparation. No matter what the result, the witness's conduct should epitomize the image of the fire service professional.

Epilogue—"But Is It Fair?"

In attempting to arrive at a course of action, do not consider only the legal aspects of your decision. Typical situations needing decisions are whether to reprimand or suspend a member for violating a rule; whether to prosecute a building owner for a code violation or grant him a time extension within which to comply; and whether to grant or deny an injured member a leave or pension. Follow the example of Earl Warren, former Chief Justice of the United States who often interrupted compelling legal arguments to ask: "But is it fair?"

That question is basic to any legal system whose end is justice. Rational minds presented with the same testimony and physical evidence can reach exactly opposite yet equally reasonable conclusions. Under these circumstances justice's answer should not end there. Rather, the final inquiry should be: "Is the result fair?" Notwithstanding the current cynicism regarding government and the law, the name of the game still is fairness. Strict adherence to the law more often than not brings about a reasonable result, in the sense that the parties involved are accorded the treatment which the law permits or requires. Such results may nevertheless appear to be highly arbitrary and unjust to a fair-minded person. Most people want more than justice; they want to be treated fairly!

For example, a fire fighter's superior officers may use their authority to change his assignment from the engine house to the fire prevention bureau without consulting him; or they may transfer him to the "big house"—headquarters—when his seniority normally would let him relax in a residential station. Such a change is made without even the gesture of asking the individual's preference. "But is it fair?" The city council or board of supervisors may exercise its power to withhold a much-needed pay raise to combat the ever-rising costs of government. Again, "But is it fair?" Fire fighters might demand a 36-hour work week and a 100 percent pension for 30 years' service, while other city employees must work a 40-hour week 40 years for a 50 percent pension. Through the exercise of political pressure, fire fighters might secure their demands. But again, "Is it fair?"

Because of the proliferation of damage claims, malpractice suits, and other court cases, there is an ever-increasing reluctance by the public to offer aid to the unfortunate. There is no law which compels a doctor, paramedic, or fire fighter to stop while off-duty and render assistance to accident victims. They can legally drive on without lending a hand. Although these professionals all have been trained—often at public expense—to assist others in danger and no doubt their inner impulses would be to offer help, justice does not mandate their efforts.

Too often people curb their inclinations toward kindness because they are afraid of some untoward result. No one should let the fear of an ingrate's lawsuit deter him from the humanitarian service that he is capable of offering. In the unlikely event that a person would have to face a judge or jury of his peers, the most important question they will have to answer will be: "But is it fair?" If a person has been fair in his treatment of others, it follows that he should take a chance on their treating him fairly in return.

Review Activities

1. A. What are the two elements in an arson-related crime? To what standard of proof must they be shown?
   B. Does your community have any programs to combat the problem of arson? If so, describe one such program. If not, research the arson control program of a neighboring community and describe how a similar program could be of benefit in yours.
2. Differentiate between the meanings of "intent" and "motive." Give two examples from your experiences.
3. When does a fire fighter have the authority to arrest someone? How is the arrest made? After an arrest takes place, what further actions are required?
4. Describe what preparations should be made in anticipation of a code violation trial? How would you prepare yourself and the prosecutor for the trial date?
5. What is circumstantial evidence and how does it differ from direct evidence? Give examples in an arson case setting.
6. A. Why are the fire inspector's demeanor and appearance important?
   B. What is the fire inspector's best method of recommending safety changes to an owner of a building?
   C. Outline the steps taken if a building owner does not comply with a fire inspector's "Recommendation Memorandum."

Table of Cases

Chapter One

State v. Sawn, 1 N.D. 5, 44 N.W. 492
U.S. Fidelity and Guarantee Co. v. Guenther, 1930, 281 U.S. 34

Chapter Three

Commonwealth v. Ritter, Court of Oyer and Terminer, Philadelphia, 1930, 13 D. & C. 285

Chapter Four

Marbury v. Madison, 5 U.S. (1 Cranch) 137, 2 L.Ed. 60 (1803)

Chapter Five

Abood v. Detroit Board of Education, 431 U.S. 209, 230 (1977)
Acevedo v. City of North Pole, 672 P.2d 130 (Alaska 1983)
Albemarle Paper Co. v. Moody, 95 S.Ct. 2362 (1975)
Arden v. Village of Oak Lawn, 537 F.Supp. 181 (N.D. Ill. 1982)
Ayres v. Indian Heights Volunteer Fire Department, 482 N.E.2d 732, 738 (Ind.App. 2 Dist. 1985) citing Indiana Code 36-8 13-2 (Burns Code Ed.1981)
Barry v. City of New York, 712 F.2d 1554 (2nd Cir. 1983)
Berkman v. City of New York, 536 F.Supp. 177, 205 (1982) citing EEOC's Uniform Guidelines on Employee Selection Procedures, 29 C.F.R. 1607 (1978); also 536 F.Supp. 177 (E.D.N.Y. 1982), aff'd 705 F.2d 584 (2nd Cir. 1983); and 580 F.Supp. 226 (E.D.N.Y. 1983)

Bickel v. Burkhart, 632 F.2d 1251, 1256 (5th Cir. 1980)
Booth v. Township of Winslow, 475 A.2d 644 (N.J.Super.A.D. 1984)
Bowman v. Pulaski County Special School District, 723 F.2d 640, 644 (8th Cir. 1983)
Brockell v. Norton, 688 F.2d 588, 593 (8th Cir. 1982); 732 F.2d 664, 667 (8th Cir. 1984)
Brown v. City of Meridian, 370 So.2d 1355, (Miss. 1979)
Calfapietra v. Walsh, 294 N.Y. 867, 62 N.E.2d 490 (1945)
Carter v. Gallagher, 452 F.2d 315 (1972)
City of Alexandria v. Alexandria Fire Fighters Association, Local No. 540, 220 La. 754, 57 So.2d 673 (La. 1952)
City of Atlanta v. Myers, 240 Ga. 261, 240 S.E.2d 60 (Ga. 1977)
City of Crowley Firemen v. City of Crowley, 280 So.2d 987 (La. 1973)
City of Erie v. International Association of Fire Fighters, Local 293, 459 A.2d 1320 (Pa.Cmnwlth. 1983)
City of Flagstaff v. Atchison, Topeka & Santa Fe, 719 F.2d 322, 325 (1983)
City of Jacksonville v. Jacksonville Association of Fire Fighters, 365 So.2nd 1098 (Fla.App. 1979)
City of Meadville Firemen's Civil Service Commission v. Neff, 450 A.2d 1078, 1080 (Pa.Cmnwlth. 1982)
City of New Rochelle v. Joseph R. Crowley, 61 A.D.2d 1031, 403 N.Y.S.2d 100 (1978)
City of St. Augustine v. Professional Fire Fighters of St. Augustine, Local 2282, IAFF, PFFF, 440 So.2d 416 (Fla.App. 1983)
Civil Service Commission of Pittsburgh v. Parks, 471 A.2d 154 (Pa.Cmwlth 1984)
Clark County v. City of Las Vegas, 574 P.2d 1013 (Nev. 1978)
Cook by Cook v. City of Geneva, 127 Misc.2d 261, 485 N.Y.S.2d 497 (Sup. 1985)
County of Los Angeles v. Davis, 440 U.S. 625 (1979)
Crain v. Firemen's and Policemen's Civil Service Commission, 495 S.W.2d 20 (Tex.Civ.App. 1973)
Davis v. County of Los Angeles, 566 F.2d 1334, 1341 (9th Cir. 1977), action held moot in *County of Los Angeles v. Davis*
Detroit Fire Fighters Association, Local 334, IAFF v. City of Detroit, 508 F.Supp. 172, 176 (E.D.Mich. 1981)
District of Columbia v. Air Florida, Inc., 750 F.2d 1077, 1080 (1984)
Driscoll v. Department of Fire of the City of Syracuse, 89 A.D.2d 1056, 454 N.Y.S.2d 563 (1982)
Ebright v. City of Whitehall, 455 N.E.2d 1307 (Ohio App. 1982)
Evangelista v. City of Rochester, 580 F.Supp. 1556 (W.D.N.Y. 1984)
Fagiano v. Police Board of City of Chicago, 456 N.E.2d 27, 29 (Ill. 1983)
Firefighters Institute for Racial Equality v. City of St. Louis, Missouri, 616 F.2d 350 (8th Cir. 1980)
Fish v. McGann, 205 Ill. 179, 68 N.E. 761 (1903)
Franklin County Sheriff's Office v. Sellers, 646 P.2d 113 (Wash. 1982)
Gailband v. Christian, 56 N.Y.2d 890, 438 N.E.2d 1116 (1982)
Gantz v. City of Detroit, 392 Mich. 348, 220 N.W.2d 433 (1974)
Garcia v. San Antonio Metropolitan Transit Authority, 105 S.Ct. 1005, 83 L.Ed.2d 1016 (1985)
Germann v. City of Kansas City, 776 F.2d 761 (8th Cir. 1985)
Givhan v. Western Line Consolidated School District, 439 U.S. 410, 415 (1979)
Glenview Rural Fire Protection District v. Raymond, 19 Ill.App.3d 272, 311 N.E.2d 302, 304 (1974)
Grady v. Blair, 529 F.Supp. 370, 372 (N.D.Ill. 1981)

Hameetman v. City of Chicago, 776 F.2d 636 (7th Cir. 1985)
Harmony Volunteer Fire Company and Relief Association v. Pennsylvania Human Relations Commission, 459 A.2d 439 (Pa.Cmnwlth. 1983)
Huhnke v. Wischer, 72 N.W.2d 915, 918 (Wis. 1955)
Hunt v. City of Peoria, 30 Ill.2d 230, 195 N.E.2d 719 (Ill. 1964)
Hutchinson v. Hughart, 285 S.E.2d 897 (W.Va. 1982)
Hutchinson v. Magee, 279 Pa.119, 120, 122 A. 234 (1923)
IAFF Local 2287 v. City of Montpelier, 133 Vt.175, 332 A.2d 79 (1975)
International Association of Fire Fighters, Local 314 v. City of Salem, 684 P.2d 605 (Ore.App. 1984)
Ittig v. Huntington Manor Volunteer Fire Department Inc., 95 A.D.2d 829, 463 N.Y.S.2d 870 (1983)
Jackson v. Byrne, 738 F.2d 1443, 1447 (1984)
James v. Prince George's County, 418 A.2d 1173, 1178 (1980)
Janusaitis v. Middlebury Volunteer Fire Department, 607 F.2d 17 (2nd Cir. 1979)
Kalamazoo Police Supervisors' Association v. City of Kalamazoo, 343 N.W.2d 601, 605 (Mich.App. 1983)
Kane v. Walsh, 295 N.Y.198, 66 N.E.2d 53 (1946)
Kelly v. Johnson, 425 U.S.238 (1976)
Klein v. Civil Service Commission, 260 Iowa 1147, 152 N.W.2d 195 (Iowa 1968)
Krolick v. Lowery, 32 A.D.2d 317, 302 N.Y.S.2d 109, 114 (1969)
McCarthy v. Philadelphia Civil Service Commission, 424 U.S. 645 (1976)
McCrea v. Cunningham, 277 N.W.2d 52 (Neb. 1979)
McDonnell Douglas Corp. v. Green, 411 U.S. 792 (1973)
Magill v. Lynch, 560 F.2d 22 (1st Cir. 1977), cert. denied, 434 U.S. 1063 (1978)
Michini v. Rizzo, 379 F.Supp. 837 (1974) affirm. 511 F.2d 1394 (3rd Cir. 1975)
Muscare v. Quinn, 614 F.2d 577 (7th Cir. 1980)
National League of Cities v. Usery, 426 U.S. 833 (1976)
New Orleans Fire Fighters Association, Local 632 v. New Orleans, 230 So.2d 326 (Ct.App. 1970) aff'd 255 La. 557, 232 So.2d 78 (1970)
Newman v. Department of Fire, 425 So.2d 753 (La. 1983)
Nick v. Department of Fire, 416 So.2d 131 (La.App. 1982)
Northcutt v. Hardebeck, 52 S.W.2d 901 (1932)
Ocean City-Wright Fire Control District v. Ocean City-Wright Fire Fighters Association, Local 2879, IAFF, 440 So.2d 413 (Fla.App. 1983)
O'Leary v. Town Manager of Arlington, 356 N.E.2d 697 (Mass. 1976)
People ex rel. Curtain v. Heizer, 36 Ill.2d 438, 223 N.E.2d 128 (1967)
People v. Edge Town Motel Corp., 26 Ill.2d 83, 186 N.E.2d 33 (1962)
People v. Wilson, 240 Cal.App.2d 574, 49 Cal.Rptr. 792 (1966)
Phillips v. Hall, 113 Ill.App.3d 409, 447 N.E.2d 418 (1983)
Pickering v. Board of Education, 391 U.S. 568
Pienta v. Village of Schaumburg, Illinois, 710 F.2d 1258 (7th Cir. 1983)
President of Williams College v. Williamstown, 219 Mass. 46, 106 N.E. 687, 688 (1914)
Rockwood Volunteer Fire Department v. Town of Kossuth, 260 Wis. 331, 50 N.W.2d 913, 914 (1952); see companion case, *Rudolph Volunteer Fire Department v. Town of Rudolph*
Rudolph Volunteer Fire Department v. Town of Rudolph, 260 Wis. 362, 50 N.W.2d 915 (1952)
Save Our Fire Department Paramedics Committee v. City of Appleton, 389 N.W.2d 43 (Wis.App. 1986)
Seyer v. Schoen, 6 A.D.2d 177, 175 N.Y.S.2d 1003, 1008 (1958)

Shelby Township Fire Department v. Shields, 320 N.W.2d 306 (Mich.App. 1982)
Smith v. City of Newark, 128 N.J.Sup. 417, 320 A.2d 212 (Law Div. 1974)
State ex inf. Gavin v. Gill, 688 S.W.2d 370, 372 (Mo.banc 1985)
State ex rel. Crites v. West, 509 S.W.2d 482, 484 (1974)
State ex rel. Kennedy v. Rammers, 340 Mo. 126, 101 S.W.2d 70 (1936)
Terry v. Bender, 143 C.A.2d 193, 300 P.2d 119 (1956)
Town of Burlington v. Labor Relations Commission, 17 Mass.App. 402, 459 N.E.2d 125 (1984)
Town of Winchester v. Connecticut State Board of Labor Relations, 402 A.2d 332 (Conn. 1978)
Travis v. City of Memphis Civil Service Commission, 671 S.W.2d 849 (Tenn.App. 1984)
United States Civil Service Commission v. National Association of Letter Carriers, 413 U.S. 548, 565 (1973)
Vair v. City of Ravenna, 29 Ohio St.2d 135, 279 N.E.2d 884 (1972)
Wachmans v. City of Dallas, 704 F.2d 160 (5th Cir. 1983)
Warner v. City of Los Angeles, 42 Cal.Rptr. 502, 231 Cal.App.2d 904 (1965)
Waters v. Chaffin, 684 F.2d 833, 836 (11th Cir. 1982) citing *Pickering v. Board of Education* and *Givhan v. Western Line School District*
Weisenritter v. Board of Fire and Police Commissioners of Burbank, 67 Ill.App.3d 799, 385 N.E.2d 336 (1979)
Whirlpool Corp. v. Marshall, 445 U.S. 11; 100 S.Ct. 883 (1980)
White v. City of Winnfield Fire Department, 384 So.2d 471 (La.App. 1980), June 12, 1980
Ziegler v. Department of Fire, 426 So.2d 311 (La.App. 1983)

Chapter Six

Almanza v. City of Chicago, Cook County, Ill., Cir. Ct., No. 82 L 17467, Apr. 24, 1985
Asian Consulting Laboratories, Inc. v. City of Saratoga Springs, 49 A.D.2d 981, 374 N.Y.S.2d 375 (1975)
Ayres v. Indian Heights Volunteer Fire Department, 482 N.E.2nd 732 (Ind.App.2 Dist. 1985)
Barnes v. Waco, 262 S.W.1081 (Tex.Comm.of App. 1924)
Barnum v. Rural Fire Protection Company, 24 Ariz.App.233, 537 P.2d 618 (Ariz. App. 1975)
Biloon's Electrical Service, Inc. v. City of Wilmington, 401 A.2d 636, 637, 641 (Del. Super. 1979)
Boothly v. Town of Yreka City, 117 Cal.App.643, 4 P.2d 589 (1931)
Campbell v. City of Bellevue, Washington, 85 Wash. 2d 1 (1975)
Canade, Inc. v. Town of Blue Grass, Iowa, 195 N.W.2d 734 (1972)
Chambers-Castanes v. King County, 669 P.2d 451 (Wash. 1983)
Chandler Supply Co. v. City of Boise, 104 Idaho 480, 660 P.2d 1323 (Idaho 1983)
City and County of San Francisco v. Superior Court, California, 160 Cal. App.3d 842 -1984)
City of Atlanta v. Roberts, Georgia, 211 S.E.2d 615 (1974)
City of Daytona Beach v. Palmer, 469 So.2d 121, 124 (Fla. 1985)
City of Fairbanks v. Schiable, 375 P.2d 201 (Alaska 1962)
City of Hammond v. Cataldi, 449 N.E.2nd 1184 (1983)

City of Houston v. Shilling, 240 S.W.2d 1010 (1951)
City of Kotzebue v. McLean, 702 P.2d 1309 (Alaska 1985)
City of Pueblo v. Flanders, 225 P.2d 832 (1950)
Commercial Carrier Corp. v. Indian River County, 371 So.2d 1010 (1979)
Coxson v. Gerry Volunteer Fire Department, Inc., 470 N.Y.S.2d 222, 98 App.Div.2d 983 (1983)
Cross v. City of Kansas City, 230 Kan.545, 638 P.2d 933 (1982)
DeFrancesco v. Western Pennsylvania Water, 478 A.2d 1295 (Pa.Super. 1984)
Delong v. Erie County, 89 A.D.2d 376, 455 N.Y.S.2d 887 (1982)
Denham v. Sears, 386 So.2d 687 (La.App. 1980)
Denton v. Page, 683 S.W.2d 180 (Tex.App. 1985), rev.29 Tex.Sup.Ct. 127 (Jan. 8, 1986)
Dougherty v. Torrence, 2 Ohio St.3d 69, 442 N.E.2d 1295 (1982)
Dougherty v. Town of Belmont, 396 Mass.271, 485 N.E.2d 183 (1985)
Evangelical United Brethren Church v. State, Washington, 407 P.2d 440 (1965)
Florio v. Schmolz, 101 N.J.L. (16 Gummere) 535, 129 A. 470; quoted with apporoval by *Ryan v. State*
Ford v. City of Caldwell, 321 P.2d 589 (Idaho 1958)
Gans Tire Sales Co., Inc. v. City of Chelsea, 16 Mass.App. 947 (1983)
Gilewski v. Mastic Beach Fire District No. 1, New York, 241 N.Y.S.2d 874 (1963)
Gilmour v. Green Village Fire Department, 2 N.J.Super.393, 63 A.2d 918 (Ch.Div. 1949)
Grant v. Petronella, California, 50, 123 Cal. App. 3d 281 (1975)
Green v. City of Knoxville, 642 S.W.2d 431 (C.A. E.D. 1982) per. for app. den. Nov. 22, 1982
Green v. Millsboro Fire Company, Inc., Delaware, 403 A.2d 286 (1979)
Harris v. Board of Water and Sewer Commissioners of the City of Mobile, 320 So.2d 624 (Ala. 1975)
Heieck and Moran v. City of Modesto, 64 Cal.2d 229, 411, P.2d 105 (1966)
Helman v. County of Warren, New York, 111 A.D.2d 560 (1985); 67 N.Y.2d 799 (1986)
Hilstrom v. City of Saint Paul, 134 Minn. 391, 214 N.W. 656 (1927)
Irizarry v. City of New York, 52 A.D.2d 807, 383 N.Y.S.2d 337 (1976)
Jackson v. Byrne, 738 F.2d 1443, 1446, 1447, 1448 (7th Cir. 1984)
Jolly v. Insurance Co. of North America, 331 So.2d 368 (Fla.App. 1976)
Karczmarczyk v. Quinn, 200 A.2d 461, 463 (R.I. 1964)
King v. City of Seattle, 84 Wash.2d 239, 525 P.2d 228 (Wash.Super. 1974)
Lewis v. Mendocino Fire Protection District, 190 Cal. Rptr. 883 (1983)
Long v. Birmingham, 49 S. 881 (1909)
Lorschbough v. Twp. of Buzzle, Minnesota, 258 N.W.2d 96 (1977)
Lucas v. Los Angeles, California, 10 Cal.App.2d 276, 75 P.2d 599 (1938)
McCurry v. City of Farmington, New Mexico, 643 P.2d 292 (1982)
Malhame v. Borough of Demarest, 162 N.J.Super.248, 392 A.2d 652 (1978)
Matlock v. New Hyde Park Fire District, 228 N.Y.S.2d 894 (1962)
Mulcairns v. Jonesville, 67 Wis.24, 29 N.W.565 (1886)
Nashville Trust Co. v. Nashville, 182 Tenn.545, 549, 188 S.W.2d 342 (1945), cited with approval in *Reynolds Boat Co. v. Haverhill*
Nicastro v. Chicago, 175 Ill.App.634 (1912)
Niehaus Brothers Co. v. Contra Costa Water Co., 159 Cal.305, 113 P.375 (1911)
Paducah Lumber Co. v. Paducah Water Supply Co., 89 Ky.340, 12 S.W.554 (1889)
Potter v. City of Oceanside, 114 Cal. App.3d 399 (1981)
Powell v. Village of Fenton, 240 Mich.94, 214 N.W.968 (1927)

Printed Terry Finishing Co., Inc. v. City of Lebanon, 372 A.2d 460 (Pa.Super. 1977)
Reynolds Boat Co. v. Haverhill, 357 Mass.668, 670, 260 N.E.2d 176 (1970)
Rowland v. Christian, 70 Cal.Rptr.97, 443 P.2d 561 (1968)
Ryan v. State, Arizona, 656 P.2d 597 (1982)
Sans v. Ramsey Golf and Country Club, 29 N.J.438, 149 A.2d 599 (1959)
Shawnee Twp. Fire District No. 1 v. Morgan, Kansas, 559 P. 2d 1141 (1977)
Sheridan Acres Water Co. v. Douglas County, 688 P.2d 297 (Nev. 1984)
Shockey v. City of Oklahoma City, 632 P.2d 406 (Okla. 1981)
State of Wisconsin v. Schoenheide, 104 Wis.2d 114, 310 N.W.2d 650 (1981)
Steele v. City of Houston, 603 S.W.2d 786 (Tex. 1980), red. den. Sept. 12, 1980, 18 U.S.C.A. Section 2102(a)
Steitz v. Beacon, 295 N.Y.51, 64 N.E.2d 704 (1945)
Thompson v. City of New York, 60 N.Y.2d 948, 459 N.E.2d 159 (1983)
Torres v. Los Angeles, 58 Cal.2d 35, 372 P.2d 906 (1962)
Trezzi v. City of Detroit, 120 Mich.App.506, 328 N.W.2d 70 (Mich.App. 1983)
Tuthill v. Rochester, 292 N.Y.S.2d 249 (1968)
Ukiah v. Ukiah Water and Improvement Co., 142 Cal.173, 75 P. 773 (1904)
Utica Mutual Insurance Co. v. Gaithersburg-Washington Grove Fire Department, Inc., 455 A.2d 987 (Md.App. 1983)
Veach v. Phoenix, 102 Ariz. 195, 427 P.2d 337 (1967)
Walters v. City of Carthage, 36 S.D.11, 153 N.W.881 (1915)
Weiss v. Fote, 167 N.E.2d 63 (1960)
Westminster Investing Corp. v. G.C. Murphy Co., 140 U.S.App.D.C.247, 434 F.2d 521 (D.C.Cir. 1970)
Williams v. City of Tuscumbia, 426 So.2d 824 (Ala. 1983)
Wilson v. City of Mason, 190 Ill.App.510 (1914)
Wong v. City of Miami, 237 So.2d 132 (1970)
Workman v. New York, 179 U.S.552 (1899)

Chapter Seven

Barger v. Mayor and City Council of Baltimore, 616 F.2d 730 (1980)
Blackmore v. Auer, 187 Kan. 434, 357 P.2d 765 (1960)
Jackson v. City of Kansas City, 235 Kan.278, 680 P.2d. 877 (Kan. 1984)
Lawrence v. City of New York, 82 A.D.2d 485, 447 N.Y.S.2d 506, 513 (1981)
Loughlin v. State of New York, 105 N.Y.159, 162, 11 N.E. 371 (1887)
Smith v. F.W.D. Corp., 436 N.E.2d 35, 37 (1982)
Taylor v. Hostetler, 186 Kan.788, 352 P.2d 1042 (1960)
Wolf v. City of New York, 39 N.Y.2d 572, 573, 384 N.Y.S.2d 758, 760 (1976)

Chapter Eight

Adams v. State, 555 P.2d 235 (Alaska 1976)
Aetna Insurance Co. v. 3 Oaks Wrecking and Lumber Co., 382 N.E.2d 283 (Ill.App. 1978); and at 288 citing *Hahn v. Eastern Illinois Office Equipment Co.*
Baltimore Planning Commission v. Victor Development Co. Inc., 261 Md.387, 275A.2d 478 (1971)
Boston and Maine Rail Road v. Armburg, 285 U.S. 234, 52 S.Ct. 336 (1932)

Jones v. United States, 362 U.S. 257, 263, 80 S.Ct. 725, 732, 4 L.Ed.2d 697 (1960)
Katz v. United States, 389 U.S. 347, 88 S.Ct. 507, 19 L.Ed. 576 (1967)
Kaufman v. Stein, 138 Ind.49, 37 N.E. 333 (Ind. 1893)
King County v. Primeau, 654 P.2d 1199 (Wash. 1982)
Land v. Mt. Vernon, 38 N.Y.2d 344, 379 N.Y.S.2d 798, 342 N.E.2d 571 (1976)
Lawton v. Steele, 145 U.S. 133 (1894)
Lorschbough v. Twp. of Buzzle, 258 N.W.2d 96 (Minn. 1977)
Luxen v. Holiday Inns, Inc., 566 F.Supp. 1484, 1486 (1983)
McCallin v. Walsh, 64 A.D.2nd 46, 407 N.Y.S.2d 852, 855, and 859 (1978) citing *Queenside Hills Co. v. Saxl*
McCulloch v. Maryland, 17 U.S. 579, 4 Wheat 316 (1819)
Maguire v. Reardon, 41 Colo.App. 596, 183 P. 303, 255 U.S. 271 (1921)
Mapp v. Ohio, 367 U.S. 643, 81 S.Ct. 1684, 6 L.Ed.2d 1081 (1961), citing *Boyd v. United States*
Mars v. Whistler, 49 Cal.App. 364, 193 P.600 (1920)
Marshall v. Barlow's Inc., 436, U.S. 307, 320, 98 S.Ct. 1816, 1824, 56 L.Ed.2d 305 (1978)
Marshall v. Kansas City, 355 S.W.2d 877 (1962)
Michigan v. Clifford, 464 U.S. 287, 104 S.Ct. 56, 59, L.Ed. 2d 486, 498, 641, 78 L.Ed. 2d 477 (1984)
Michigan v. Tyler, 436 U.S. 499,510, 98 S.Ct. 1942,1959, 56 L.Ed.2d 486,498 (1978)
In re Mingo, 190 Cal.769, 214 P.850 (1923)
Moch v. Rensselaer Water Co. 247 N.Y.160, 159 N.E.896 (1928)
Morris v. Marin, 18 Cal.3d 901, 136 Cal.Rptr. 251, 559 P.2d 606 (1977)
In re M/T Alva Cape, 405 F.2d 962, 968 (2nd Cir. 1969)
National Tank Truck Carriers, Inc. v. Burke, 535 F.Supp. 509 (D.C. R.I. 1982), affirmed 698 F.2d 559 (CA1 1982)
National Tank Truck Carriers, Inc. v. New York, 677 F.2d 270 (CA2 1982)
Parker v. Kirkwood, 134 Kan.749, 8 P.2d 340 (1932)
People v. Miller, 176 N.Y.S.206 (1917)
Queenside Hills Co. v. Saxl, 328 U.S. 80, 66 S.Ct. 850, 90 L.Ed. 1096
Quinn V. Nadler Brothers, Inc., 59 N.Y.2nd 914, 466 N.Y.S.2d 292, 453 N.E.2d 521 (1983)
Ribeiro v. Town of Granby, 395 Mass.608, 481 N.E.2d 466 (Mass. 1985) citing *Irwin v. Ware*
Rowland v. Christian, 69 Cal.2d 108, 70 Cal.Rptr. 97, 443 P.2d 561 (1968)
Roxas v. Cogna, 41 Cal.App.2d 234 (1940)
Sanchez v. Village of Liberty, 42 N.Y.2d 276, 366 N.E.2d 870 (1977)
Schwartzmann v. Merritt Island Volunteer Fire Department, 352 So.2d 1230 (Fla.App. 1977)
See v. City of Seattle, 387 U.S. 541, 87 S.Ct. 1737 (1967)
Siple v. City of Topeka, 679 P.2d 190 (Kan. 1984)
Smith v. F.W.D. Corp., 436 N.E.2nd 35 (Ill.App. 1982)
Smuullen v. City of New York, 28 N.Y.2d 66, 268 N.E.2d 763 (1971)
South Dakota v. Jorgensen, 333 N.W.2d 725 (S.D. 1983) citing *Michigan v. Tyler*
State ex re Brook v. Cook, 84 Mont. 478, 276 P. 958 (1929)
State v. Zindros, 189 Conn.228, 456 A.2d 288 (Conn. 1983)
Stewart v. Schmeider, 386 So.2d 1351, 1358 (La. 1980)
Thorpe v. Rutland Railroad Co/m5./m1, 27 Vt. 140 (1855)
Trianon Park Condominium Association, Inc. v. City of Hialeah, 468 So.2d 912, 920, 922 (Fla. 1985)
United States v. Calandra, 414 U.S. 338,347, 94 S.Ct. 613,619, 38 L.Ed.2d 561 (1974)

Westchester Rockland Newspapers Inc. v. Kimball, 50 N.Y.2d 575, 408 N.E.2d 904 (1980)
Williams v. City of North Las Vegas, 91 Nev.622, 541 P.2d 652 (1975)
Yalowizer v. Husky Oil Co., 629 P.2d 465 (Wyo. 1981)

Chapter Nine

Atchley v. City of Fresno, 199 Cal.Rptr. 72, 151 Cal.App.3d 635 (1984)
Baker v. Fire Pension Fund Commissioners, 18 Cal.App. 433, 123 P. 344 (1912)
Ballard v. Board of Trustees of the Police Pension Fund of the City of Evansville, 263 Ind. 79, 324 N.E.2d 813 (1975)
Barber v. Jewel Tea Co., 252 App.Div. 362, 300 N.Y.S. 302 (1937)
Board of Trustees of the Oklahoma City Police Pension and Retirement Plan v. Clark, 661 P.2d 506, 509 (Okla. 1983)
Borough of Beaver v. Liston, 464, A.2d 679 (Pa.Cmnwlth. 1983)
Carreras v. McGuire, 49 N.Y.S.2d 506 (1982 citing with approval *Barber v. Jewel Tea Co.*)
Carrich v. Sherman, 106 Cal.App.546, 228 P.143 (1930)
Elliott v. City of Boston, 245 Mass.330, 139 N.E.523 (1923)
Equal Employment Opportunity Commission v. City of Altoona, Pennsylvania, 723 F.2d 4 (1983)
Fairfax County Fire and Rescue Service v. Newman, 281 S.E.2d 897, 899 (Va. 1981)
Garcia v. San Antonio Metropolitan Transit Authority, 105 S.Ct. 1005, 83 L.Ed.2d 1016 (1985)
Hill v. City of Lincoln, 213 Neb.517, 330 N.W.2d 471 (Neb. 1983)
Johnson v. Mayor and City Council of Baltimore, 105 S.Ct. 2717 (1985)
Kern v. City of Flint, 335 N.W.2d 708, 710 (1983)
Majauskas v. Majauskas, 61 N.Y.2d 481, 474 N.Y.S.2d 699, 463 N.E.2d 15 (1984)
Meaney v. Regan, 88 A.D.2d 1020, 451 N.Y.S.2d 915 (1982)
Peterson v. Board of Trustees of the Firemen's Pension Fund, 296 N.E.2d 271 (Ill. 1973)
San Francisco Fire Fighters Local 798 v. City and County of San Francisco, 152 Cal.App.3d 113, 116, 118, 201 Cal.Rptr. 176, 178
Taggart v. City of Asbury Park, 15 N.J.Misc.10, 188 A.490 (1936)
Tampa v. Jones, 18 So.2d 1150 (Fla.App. 1984)
Traub v. Board of Retirement of the Los Angeles County Employees Retirement Association, 34 Cal.3d 793, 195 Cal.Rptr. 681, 670 P.2d 335 (Cal. 1983)
Wells v. Police and Firefighters Retirement and Relief Board, 459 A.2d 136 (D.C.App. 1983)
Winslow v. City of Pasadena, 34 Cal.3d 66, 192 Cal.Rptr. 629, 665 P.2d 1, 3 (Cal. 1983)

Chapter Ten

Bernthal v. City of St. Paul, 376 N.W.2d 422 (Minn. 1985)
Chalachan v. City of Binghampton, 55 N.Y.2d 989, 434 N.E.2d 257 (1982)
Cuna v. Board of Fire Commissioners, Supreme Court of New Jersey (1964)
Doody v. Davie, 77 Cal.App. 310, 246 P. 339 (1926)

Garcia v. San Antonio Metropolitan Transit Authority, 105 S.Ct. 1005, 83 L.Ed.2d 1016 (1985)
Harleysville Mutual Insurance Co. v. Five Points Fire Co. No. 1 Inc., 444 A.2d 304 (Del.Super. 1982)
Hennige v. Fairview Fire District, 99 A.D.2d 158, 472 N.Y.S.2d 204 (1984)
Hill v. City of Camden, 463 A.2d 1982 (1983)
Keogh v. City of Bridgeport, 444 A.2d 225, 230-232 (Conn. 1982)
Lamalfa v. Denny, 34 A.D.2d 709, 309 N.Y.S.2d 852 (1970)
Logue v. Carthage, 612 S.W.2d 148 (Mo.App. 1978)
Maines v. Cronomer Valley Fire Department, Inc., 429 N.Y.S.2d, 407 N.E.2d 466 (1980)
Midwest City v. Cravens, 532 P.2d 829 (1975)
National League of Cities v. Usery, 426 U.S. 833 (1976)
Office of Emergency Services v. Home Insurance Co., 618 S.W.2d 573 (Ark. 1981)
Padden v. New York City, 92 N.Y.S 926, 45 Misc.Rep. 517 (1904)
Rogers v. Town of Newton, 433 A.2d 1303 (N.H. 1981)
St. Louis Fire Fighters Association, Local 73, AFL-CIO v. City of St. Louis, 637 S.W.2d 128 (Mo.App. 1982)
Village of Creve Coeur v. Industrial Commission, 32 Ill.2d 430, 206 N.E.2d 706 (1965)

Chapter Eleven

Aiello v. City of Wilmington, 623 F.2d 845 (1980)
Anabel V. Ford, Civil No. 84-6033 (W.D.Ark. Sept. 5, 1986)
Arrastia v. Lowery, 356 N.Y.S.2d 306 (1974)
Board of Regents v. Roth, 408 U.S. 564, 92 S.Ct. 2701, 33 L.Ed2d 548 (1972)
Bredeson v. Crofti 295 Ala.246, 326, So.2d 735, 737 (1976)
Cann v. Civil Service Board of Oakland, Cal.Supr.Ct. 1975; 9 Fire Dept. Pers. Rep. 7 (Sept. 1975)
Capua v. Plainfield, 643 F.Supp.1507, (D.N.J. Sept. 18, 1986)
City of Erie v. Kelley, 474 A.2d 1226 (Pa.Cmnwlth. 1984)
City of Gadsden v. Head, 429 So.2d 1005 (Ala. 1983)
City of Palm Bay v. Bauman, 475 So.2d 1322 (Fla.App. 1985)
Cleveland Board of Education v. Loudermill, 105 S.Ct. 1487 (1985)
Connick v. Myers, 461 U.S. 138, 147, 103 S.Ct. 1684, 1690, 75 L.Ed.2d 708 (1983)
Danner v. Bristol Twp. Civil Service Commission, 440 A.2d 702, 703 (Pa. Cmnwlth. 1982) citing Zeber Appeal
Debnam v. Town of Belmont, 447 N.E.2d 1237, 1240 (Mass. 1983)
Fire Fighters Local Union No. 1784 v. Stotts, 467 U.S. 561, 577, 81 L.Ed.2d 483, 497, 104 S.Ct. 2576 (1984)
Garrity v. New Jersey, 385 U.S. 493, 87 S.Ct. 616, 17 L.Ed.2d 562 (1967)
Garvin v. Chambers et al., 195 Cal.212, 232 P.696 (1924)
Jones v. McKenzie, 628 F.Supp. 1500 (D.D.C. 1986)
Kaske v. City of Rockford, 96 Ill.2d 298, 450 N.E.2d 314 (Ill. 1983)
Krolick v. Lowery, 35 A.D.317 (1980)
Leis, 455 A.2d 1277 (Pa.Cmnwlth. 1983)
Local 93 v. City of Cleveland, 106 S.Ct. 3063, 3072 (1986)
Long Beach City Employees Association v. City of Long Beach, 41 Cal.3d 937, 719 P.2d 660 (Cal. 1986)

Louisville Professional Fire Fighters Association v. City of Louisville, 508 S.W.2d 42 (Ky. 1974)
McCleod v. City of Detroit, 39 F.E.P. Cases 225 (E.D.Mich. 1985)
McCloud v. Whitt, 639 S.W.2D 375 (Ky.App. 1982)
McDonald v. Miller, 590 F.2d 1099 (1979)
Miller v. Town of Batesburg, 257 S.E.2d 159, 160 (S.C. 1979)
Murphy v. Houston, 250 Ill.App.385 (1928)
Nelson v. Baker, 112 Ore.79, 277 P.301, 303 (1924)
Oberg v. City of Billings, 674 P.2d 494 (Mont. 1983)
Ricca v. City of Baton Rouge, 450 So.2d 1032, 1034 (La.App. 1984)
Rivera v. City of Douglas, 644 P.2d 271, 273 (Ariz.App. 1982)
Schwab v. Bowen, 363 N.Y.S.2d 434 (1975)
Shelby Twp. Fire Department v. Shields, 320 N.W.2d 306 (Mich.App. 1982)
Stanfill v. City of Fairbanks, 659 P.2d 579 (Alaska 1983)
State v. Barry, 123 Ohio St.458, 175 N.E.855 (1931)
State ex rel. Kennedy v. Rammers, 340 Mo.126, 101 S.W.2d 70 (1936)
Timpano v. Hanna, 335 N.Y.S.2d 226 (1974)
Travis v. City of Memphis Civil Service Commission, 671 S.W.2d 849 (Tenn.App. 1984)
Turner v. Fraternal Order of Police, 500 A.2d 1005 (D.C.App. 1985)
Williams v. City of Montgomery, 550 Fed.Supp.662 (1982)
Wood v. City Civil Service Commission, 45 Cal.App.3d 105, 119 Cal.Rptr. 175 (1975)
Zeber Appeal, 398 Pa. 35, 156 A.2d 821, 825 (1959)

Chapter Twelve

Armstrong V. Mailand, 284 N.W.2d. 343 (Minn. 1979)
Bartels v. Continental Oil Co., 384 S.W.2d 667 (Mo. 1964)
Buchanan v. Prickett and Son, Inc., 203 Neb. 684, 279 N.W.2d 855, 858, 859 (Neb. 1979)
Buckeye Cotton Oil Co. v. Campagna, 146 Tenn.389, 212 S.W.646 (1922)
Dini v. Naiditch, 20 Ill.2d 406, 170 N.E.2d 881 (1960)
Donnelly v. City of New York, Bronx County Supreme Court, Index No. 14834/79, March 6, 1986
Houston Belt and Terminal Railway v. O'Leary, 136 S.W.601 (Tex.Civ.App. 1911)
Koehn v. Devereaux, 495 N.E.2d 211, 215 (Ind.App. 1986)
Krauth v. Geller, 31 N.J.270, 157 A.2d 129 (1960)
Lipson v. Superior Court, 31 Cal.3d 362, 371, 644 P.2d 822, 828 (1982)
Luetje v. Corsini, 126 Ill.App.3d 74, 466 N.E.2d 1304, 1305, 1306 (1984)
Lunt v. Post Printing and Publishing Co., 48 Colo.316, 110 P.203 (1910)
McLaughlin v. Rova Farms Inc., 56 N.J.288, 305, 266 A.2d 284 (1970)
Mahoney v. Carus Chemical Co., Inc., 102 N.J.564, 510 A.2d 4, 8, 9 (1986) citing *Krauth v. Geller*
Moreno v. Marrs, 102 N.M.373, 695, P.2d 1322 (N.M.App. 1984)
Pennebacker v. San Joaquin Light and Power Co., 158 Cal.579, 112 P.459 (1910)
Rowland v. Christian, 70 Cal.Rptr.97, 443 P.2d 561, 568 (1968)
Rowland v. Shell Oil Co., 224 Cal.Rptr.547, 549 (1986)
Verges v. New Orleans Public Service, Inc., 445 So.2d 878 (La.App. 1984)
Wagner v. International Railway Co., 232 N.Y.176, 133 N.E.437 (1921)
Walsh v. Madison Park Properties, Ltd., 102 N.J.Super.134, 245 A.2d 512 (1968)

Walters v. Sloan, 20 Cal.3d 199, 571 P.2d 609 (1977)
Washington v. Atlantic Richfield Co., 66 Ill.2d 103, 109, 361 N.E.2d 282, 285 (1976)
Whitten v. Miami-Dade Water and Sewer Authority, 357 So.2d 430 (Fla.App. 1978), cert. den. 364 So.2d 894

Chapter Thirteen

Archie v. City of Racine, 627 F.Supp. 766 (E.D.Wis. 1986)
Berger v. City of University City, 676 S.W.2d 39 (Mo.App. 1984)
City of Dover v. International Association of Fire Fighters Local 1312, 322 A.2d 918 (N.H. 1974)
Dalehite v. United States, 346 U.S. 15, 36 (1953)
Doeg v. Cook, 126 Cal.213, 58 P.707 (1899)
Fulenwider v. Fire Fighters Association Local Union 1784, 649 S.W.2d 268 (Tenn. 1982)
Hamilton v. Rodgers, 783 F.2d 1306, 1308 (5th Cir. 1986)
Heimberger v. City of Fairfield, 117 Cal.Rptr. 482 (Cal.App. 1974)
King v. Williams, 5 0hio St.3rd, 137, 449 N.E.2nd, 452 (1983)
Miller v. Mountain Valley Ambulance Service, Inc., 694 P.2d, 364 (Colo.App. 1984)
Nearing v. Weaver, 295 Ore.702, 670 P.2d 137, 143 (Ore. 1983)
Owen v. City of Independence, Missouri, 445 U.S. 622, 647 (1980)
Pattern Makers League of North America v. National Labor Relations Board, 724 F.2d 57, affirmed June 27, 1985
Pembaur v. City of Cincinnati, 84-1160, 8 Supreme Court Bulletin 11, March 25, 1986
Raynor v. Arcata, 11 Cal.2d 113, 121, 77 P.2d 1054 (1938)
State of Missouri v. Kansas City Fire Fighters Local 42, 672 S.W.2d 99, 109, 124 (Mo.App. 1984)
Taplin v. Town of Chatham, 390 Mass.1, 453 N.E.2d 421 (Mass. 1983)
Watson v. McGee, 527 F.Supp. 234 (1981)
In re Webb, 52 Cal.App.3d 648 (1975)
Wheeler v. City of Pleasant Grove, 664 F.2d 99 (5th Cir. 1981)

Chapter Fourteen

Baker v. State, 120 Wis.135, 97 N.W.2d 566 (1903)
Bragg v. State, 334 S.E.2d 184 (Ga.Ct.App. 1985)
Edwards v. Arizona, 451 U.S. 477 (1981)
Illinois v. Gates, 103 S.Ct. 2317, 2332 (1983)
Miranda v. Arizona, 384 U.S. 436, 439, 86 S.Ct. 1602, 16 L.Ed.2d 694 (1966)
People v. Marin, 65 N.Y.2d 741 (N.Y. 1985)
People v. Sherman, 97 Cal.App.2d 245, 217 P.2d 715 (1950)
Russell v. United States, 738 F.2d 825 (7th Cir. 1984) affirm ______ U.S. ______, 84-435 (June 3, 1985)
State v. Baker, 195 Conn.598, 489 A.2d 1041 (Conn. 1985)
State v. Grier, 307 N.C.628, 300 S.E.2d 351 (1983)
State v. Kitowski, 44 Wis.2d 259, 170 N.W.2d 703 (1969)
State v. Nielson, 24 Utah 2d 11, 474 P.2d 725 (Utah 1970)
State v. Showaker, 721 P.2d 892 (Utah 1986)
Tennessee v. Garner, 710 F.2d 240 (6th Cir. 1984) affirm ______ U.S. ______, 83-1035 (March 27, 1985)

Glossary of Legal Terms

This glossary consists mainly of the legal terms used in this book. For further reference, many legal terms are included in any good desk dictionary, and detailed definitions appear in the standard legal dictionaries such as *Black's Law Dictionary*, found in most libraries.

abandonment As applied to property, an owner's voluntary relinquishment of possession of a thing, with the intention of terminating his ownership.
abrogate To cancel or repeal; to declare a prior law or act null and void.
absolute Unconditional; unrestricted.
absolute liability Responsibility without fault or negligence.
acknowledgment An admission or confirmation; public declaration that the act or deed named in a legal document was his.
accessory An individual who has aided in the perpetration of a crime before, during, or after its commission.
action An ordinary proceeding or suit brought before a court; a formal demand of one's rights against another, presented before a reviewing court.
actionable Remediable by an action at law.
admissible evidence Evidence which a court or tribunal may legally receive and consider in deciding a case.
admission A statement in conflict with or against the proprietary or pecuniary interest of the person making it.
affidavit A person's voluntary written or printed factual statement which is confirmed by oath before a public officer empowered to administer such oaths.
agent A representative vested with authority to create obligations for the person he represents.
alias Otherwise; "also known as."

allege To plead, recite, or charge.
amicus curiae "Friend of the court"; includes both laymen and lawyers who are requested to advise or whose opinions are sought by the court.
answer A plea posed by the defendant in response to the plaintiff's complaint.
ante-date To insert a date that is prior to the actual date of a document's execution.
appeal A demand or request to a higher court to review the actions or opinions of a lower court.
appellant A person who files or requests an appeal.
appellate jurisdiction The authority of a higher court to review the decision or opinion of a lower court.
arraignment A formal judicial proceeding whereby the accused is apprised of the charges against him and asked to plead "guilty" or "not guilty" to them.
arrest To detain a person or take him into custody. No physical restraint is necessary, for it is sufficient if the person being arrested understands that he is under the power of the arresting party and that his freedom of movement has been taken away.
arson At common law, it was the willful and malicious burning of the dwelling house of another; its meaning has been broadened by statute in many jurisdictions to include a person's house, outbuildings, appurtenances thereto, and even vehicles.
assignment The transfer of a property right from one person to another.
assumption of the risk A defense to an action of negligence whereby the plaintiff is shown to have knowingly and voluntarily assumed the risk of injury, reasonably or unreasonably.
attorney in fact A private person who has been authorized to do a particular thing, which binds the principal, or to otherwise act in his stead; a private attorney as distinguished from an attorney at law.
attractive nuisance doctrine The view that one who maintains items on his premises of a character likely to attract children in play, or who permits dangerous conditions to remain on his premises knowing that children are in the habit of playing around them, is liable to a child injured therefrom.
authority having jurisdiction The person, agency, or entity which exercise the right of direction or which can exact obedience by another party. The power for directing or exacting obedience may be granted by statute, ordinance, or administrative regulation, or agreed to by contract.
bail A guarantee made by the defendant or a third party, secured by some item or amount of money, to assure that the defendant will appear in court to answer the charges.
bona fide "Good faith"; without fraud or collusion.
bribery The offering or giving of anything of value to an individual or official party to influence his actions.

brief A detailed legal statement of a party's position on a case.
business visitor An individual invited or permitted on the land of another person in furtherance of the business relationship between them.
case A contested question in a court of justice.
cause of action The existence of facts or factors that give an individual a right to judicial relief from another person; the legal or equitable basis for the judicial remedy due.
caveat emptor "Let the buyer beware."
certiorari "To be informed of." A formal directive used by a superior court to direct production of the judicial record of an inferior court in order to review its decision; most commonly found in petitions for "Writs of Certiorari" to the United States Supreme Court, requesting its discretionary review of another court's ruling.
circumstantial evidence *See evidence.*
cite To command, as in a citation for corrective action; also to refer to legal proceedings or statutes in support of one's position.
civil action A civil suit; an action brought to enforce a civil right of the plaintiff, as distinguished from a criminal action or prosecution.
code A compilation of all laws, rules, or regulations on a subject (such as a building code or municipal code).
common law Law developed from publicly recognized usage and custom, as distinguished from law developed through legislation or the drafting of a constitution. Common law generally refers to the law developed through judicial precedent.
complainant The one who files a complaint; the plaintiff in a legal action.
complaint The claim made by the plaintiff; a short statement explaining the basis of the plaintiff's claim and saying that the court has authority to render assistance, and should do so, in the plaintiff's behalf.
compounding a crime The illegal act of receiving payment in return for a promise not to prosecute or inform on someone who has committed a crime.
condemnation Exercise of the power of eminent domain.
confession The voluntary statement or declaration by a person who has committed a crime or misdemeanor, acknowledging his guilt and the nature and extent of his wrongdoing.
confiscate To appropriate property; to transfer property, such as contraband or property of an enemy, from a private to a public use.
constitutional law The organic law of a state or nation that outlines the organization, powers, and framework of the government; it also may outline the fundamental rights and privileges of its citizens and the people's relationship to government.
constitutional right A right guaranteed to the individual by a constitution.
constructive notice Notice imputed to a person, based upon the circumstances, when he has not received actual notice.

contra "Against"; otherwise; disagreeing.
contraband Any property that is unlawful to produce or possess. Examples include smuggled goods or illegal drugs.
contract An agreement, oral or written, by which a person undertakes to do or not to do a particular thing for a consideration, such as a sum of money.
contributory negligence Negligence on the part of the plaintiff that makes the injury the result of the united, mutual, concurring, and contemporaneous negligence of both the plaintiff and the defendant; failure of the plaintiff to exercise ordinary care.
corporate function "Private" as distinguished from "public" activities of an incorporated governmental entity.
corpus delicti The "body" of the offense; the essence of the crime.
crime An act committed (or omitted) in violation of public law forbidding (or commanding) it.
criminal law Law which defines criminal offences and the penalties for their commission.
criminal negligence Acts of negligence that show a wanton or willful intent, with reckless disregard for the interests of another person.
criminal offense See *criminal law.*
decision A judgment given by a competent tribunal after consideration of all the facts and the applicable law. The "decision" should not be confused with the "opinion," which represents the reasons given by the court for its decision.
defendant The party who is defending or denying the allegations.
deposition A discovery device whereby the individual being deposed is placed under oath and requested to truthfully respond to questions; the replies are recorded and become admissible evidence in any subsequent trial.
dictum The court's expression of which technically is not part of the opinion; a statement or observation of the court.
directed verdict A verdict which the jury returns as directed by the court. The court takes over and directs the verdict where there is no competent, relevant, and material evidence to support the allegations; or where the evidence is contrary to all reasonable probabilities; or where it is clear that any other verdict would be against the obvious weight of the evidence.
due process of law The exercise of law that affords appropriate protections to individual interests and rights; the legal process that conforms with recognized standards of fairness and justice in its execution.
easement The right which one person has to use the land of another for a specific purpose and duration.
eminent domain The superior right of the government, through payment of just compensation, to take private property or control its use for the greater benefit of the public.
enjoin To command or require a person to do or not to do a particular thing; to issue an injunction.

entrapment Actions by officers of government to induce a person to commit a crime he did not originally intend to commit, in order for the government to undertake a criminal prosecution.
equity In its broadest meaning, this signifies natural justice; administration of the laws in the interest of fairness—not just rote enforcement of the rules.
estoppel A bar against a person's allegation or denial of a fact that is contrary to his previous allegation or denial of the same fact.
evidence Anything that tends to prove or disprove any matter in question, or to influence the belief respecting it. *Direct evidence* tends to immediately prove the existence of the fact or issue, as, for example, if A saw B set A's house on fire. *Circumstantial evidence* is not drawn from personal knowledge or direct observation and tends to prove other facts via deduction; for example, where A found that his house had been set on fire by oil-soaked rags which he proved came from B's premises and B, an old enemy of A, had been seen running away from A's house just before the fire broke out, a jury could reasonably infer from the indirect or circumstantial evidence presented that B set A's house on fire.
execution In civil actions, the mode of obtaining payment of the debt or damages; the procedural means to recover on a judgment.
ex parte On or from one side only; an "ex parte hearing" is where only one party is represented.
ex post facto law A law which makes an act punishable in a manner or to an extent to which the act was not punishable when it was committed; a law which makes an act a crime, although the act was not criminal at the time it was performed.
ex rel. Of or relating to; legal proceedings instituted by the state attorney general, on behalf of the state, upon instigation or request of a private party.
false arrest An unlawful physical restraint by one person on the liberty of another; it does not require the actual touching of the person arrested, but merely that he be made to understand that he is within the power of the person making the arrest.
fellow servant rule The common law doctrine that where the employer, or master, provides a safe place to work, he is not liable for an injury to one of his servants caused by another servant's use or misuse of any of the tools or appliances provided; "fellow servants" are all who serve a common master and who take the risk of the others' negligence.
felony A crime so declared by statute, or considered as such at common law, whose punishment is either death or imprisonment for more than one year.
governmental function An undertaking by a political subdivision which is essential to its existence; for example, regulation of land development and use, and operation of police departments, as distinguished from private functions such as private water and power systems and bus service. The distinction is important because the general rule is that a municipality is not liable

for injuries in a civil action arising out of the performance of a truly governmental function.

habeas corpus "You may have the body"; a writ directed to the person detaining someone, commanding him to produce the prisoner at a certain time and place, and to show cause why the prisoner should not be released.

hearing A proceeding of relative formality in which fundamental due process—that is, the opportunity to be heard and to have their concerns considered—is accorded all interested parties.

hearsay evidence A statement other than one made by the declarant while testifying at a trial or hearing, offered in evidence to prove the truth of the matter asserted; statements coming from a person who is not a participant in the proceeding or statements not made under oath, so that their content or truthfulness is questionable. Such evidence is not admissible in a court of justice.

home rule A state constitutional provision that grants a measure of self-government to a city or town.

impeach To challenge someone's credibility; to accuse him; or to charge him with liability.

imputed Attributed to a party or individual based upon his position or relationship with another; assumed to be possessed, such as "imputed knowledge."

incriminate To expose to an accusation or charge of a crime.

indictment An accusation, in writing, formulated by a grand jury.

inference A permissible deduction from the evidence.

information An accusation of a crime, made by a public officer against an individual.

injunction An order issued by the court, prohibiting a person or entity, or servants there of, from performing or continuing to perform some act.

in personam "Against the person."

in re "In the matter of."

in rem Against a thing rather than against a person; an "in rem proceeding" regards the status, ownership, etc. of a thing, not a person.

instrument A formal, written document intended to serve as evidence of some agreement, exchange, expression, or understanding.

interrogation The custodial or noncustodial propounding of questions by criminal justice personnel to the person arrested or suspected of a crime.

invitee One who comes upon the premises by the express or implied invitation of the owner.

judgment The final determination of a court of competent jurisdiction upon matters submitted to it in an action or proceeding.

judicial notice Said to be taken of facts which are deemed by their very nature to be already known to the court and jury and which therefore do not have to be proven; for example, that July 4, 1987, fell on a Saturday - a fact of which there can be no reasoned dispute.

judicial system The network of courts and other judicial bodies within a state, or nation.
jurisdiction The right of a court to adjudicate concerning the parties and subject matter in a given case. See *authority having jurisdiction.*
laches The doctrine that a individual cannot "sleep upon his rights" for an unreasonable length of time and then come into court and ask to be given a remedy. The intent of the laches doctrine is to avoid undue prejudice to the defending party by the tardy actions of the plaintiff. "Equity aids the vigilant and not those who slumber on their rights."
latent defect A hidden or concealed defect that could not be discovered during a normal inspection.
law The body of rules, regulations, and standards of action and conduct prescribed by the controlling authority and having binding force upon the individual.
lawsuit An action or proceeding in a civil court.
liability Any legal debt, responsibility, or obligation that one is or may be bound to fulfill.
libel A published expression (writing, sign, or picture) that is injurious to a person's character or reputation.
licensee A person who is licensed; a person who is privileged to enter upon the premises of another for his own purposes, with express or implied permission of the owner, or whose presence on the premises is merely tolerated.
line of duty The scope of one's responsibilities, or activities reasonably incidental to those responsibilities.
malice Evil intent; an act done with malice is intentional and without justification. Malice does not necessarily imply ill will toward a given person, but denotes a mental condition manifested by intentionally doing a wrongful act without just cause or excuse.
malicious prosecution Initiation of any legal action or proceeding against a person without probable cause and with malice.
mandamus "We command"; the remedial writ, issued by a court of jurisdiction, directing performance of some act.
master-servant relationship The relation between two persons when one of them may control the work of the other, and may direct the manner in which the work shall be done.
ministerial duty A duty so definitely prescribed by law that it leaves no room for discretion on the part of the officer upon whom the duty is imposed.
ministerial function A function of a municipal corporation which is a private "corporate" function as distinguished from a public "governmental" function; a function for which there is no occasion for the corporate body to use judgment or discretion.
misdemeanor Any crime that is punishable neither by death nor by imprisonment in a state prison; that is not a felony; and that typically is punishable by incarceration in a jail for less than one year.

municipal corporation A territorial and political subdivision of government established by a state for the purpose of administering local government; a charter or general law city.
negligence Failure to act as an ordinary, prudent, and reasonable person would act under like circumstances.
negligence *per se* "Negligence in itself"; or negligence as a matter of law; conduct which may be declared as negligent without any argument or proof of the surrounding circumstances. It typically is held negligence *per se* to violate a law designed to protect the general public.
nolo contendere "I will not contest it." A plea which a defendant in a criminal prosecution may have entered for him, and which amounts to an admission of "guilt" without his actually pleading "guilty." A *nolo contendere* plea cannot be used in any subsequent civil action, whereas a guilty plea may be considered during subsequent civil litigation.
noticed hearing A hearing of formal character wherein the time, place, nature of proceedings, and extent of action contemplated by the hearing body have been communicated to all persons or parties potentially having an interest.
nuisance Anything or any activity that results in injury to an individual or the public by causing annoyance, discomfort, or inconvenience.
nuisance *per accidens* A condition which by itself does not ordinarily constitute a nuisance, such as a house, but which if broken down, unsanitary, a fire menace, etc. might become a nuisance under certain circumstances.
nuisance *per se* Any condition which at all times and under all circumstances, irrespective of location or surroundings, amounts to a nuisance.
ordinance A rule or law established by a local government authority.
organic law See *constitutional law*.
original jurisdiction Jurisdiction in the first instance, typically referring to the first level of courts that is legally empowered to consider a case, as distinguished from a court that has appellate jurisdiction.
patent defect A defect which is clearly visible or obvious; not latent.
per se "Through itself"; as such.
plaintiff The person who brings a suit; the complaining party.
pleadings Allegations made by the parties to a civil or criminal case, whether to support or defeat the claims made.
police power The power of the state—and, as delegated, local government—to regulate the freedom and property rights of individuals for the protection of the health, safety, and welfare of the public.
preclude To prohibit or prevent an occurrence.
presumption Something which may be assumed without proof, or taken for granted; an assumption of fact resulting from a rule of law which requires a fact to be assumed from the proving of another fact. For example, the law presumes that a child born in wedlock is legitimate where there has been opportunity for access and the husband is not impotent. Presumptions may be either conclusive or rebuttable.

prima facie "Made first"; a first showing; something *prima facie* correct is presumed to be correct until the presumption is overcome by evidence which clearly rebuts it. For example, *prima facie* speed limits set by statute establish the maximum speed deemed safe for ordinary conditions, but do not preclude arrest for going at that speed where the conditions are so hazardous that a slower speed is required for safe operation.
principal In criminal law, anyone who is present at the commission of a crime and who participates in the crime either directly or indirectly.
probable cause Reasonable cause or basis; having more evidence for than against. For example, there is probable cause for arresting a person on a felony charge when circumstances are strong enough to indicate that the charge is true and that the individual perpetrated the wrong.
probation Releasing a person into the community, subject to conditions of personal conduct approved by a parole officer. Also, the initial tryout period of a new employee or of an existing employee promoted to a new position.
procedure The judicial process for enforcing a person's right; the sequence of judicial actions or events required to enforce a legal right.
promulgate To develop, publish, and announce officially; the formal act of announcing a new statute, ordinance, or judicial rule of law.
proprietary function An activity or undertaking of government that is nonessential to conducting the state-empowered obligations/authorities of local government; an activity undertaken for the specific benefit of the community's residents and not for the public as a whole.
proximate cause The cause of an injury, produced in a natural and continuous sequence, unbroken by any efficient intervening cause, and without which the injury would not have occurred.
public office A position specifically created by law, for a fixed term, in which the office holder is empowered to exercise specified sovereign powers of government.
public officer An individual holding public office, as distinct from a public employee; the office holder is empowered to undertake certain discretionary activities of government.
quasi "As if"; relating to or having the character of.
quasi judicial The acts of an officer which are executive or administrative in character and which call for exercise of the officer's judgment and discretion. For example, a hearing officer, while deliberating upon disciplinary charges lodged against a fire fighter, is conducting a *quasi judicial* activity.
quo warranto "By what authority." The procedure whereby the right of a person to hold an office can be contested.
removal Dismissal of an individual from a position or office by act of a competent governmental body or official.
res ispa loquitur "The thing speaks for itself." A rebuttable presumption that the defendant was negligent, based upon proof that the activity or in-

strumentality was in his exclusive control, and that any injury therefrom would not have occurred but for someone's negligence.

res judicata An issue, fact, or matter between specified parties which has been definitely settled by decision of a court; such a decision is conclusive regarding the original or any subsequent litigation.

respondeat superior "Let the superior respond"; let the master be responsible for the acts of his servant, or the principal for the acts of his agent. The doctrine requires the individual with higher authority to be responsible for the actions of the person with less authority.

respondent The party who is adverse to the appellant in an action which has been appealed to a higher court.

retrospective law A law that contemplates prior activity; one which takes away or impairs vested rights acquired under existing law, or creates a new obligation or duty regarding past transactions. A retroactive law.

rule An established or prescribed standard, guide, or regulation.

search An examination of an individual's person, property, papers, or effects for contraband or evidence of a crime.

seizure The act of taking possession of contraband, of evidence of a crime, or of a person. Seizure has occurred when the item has been possessed or movement of the individual has been restricted.

stare decisis "Let the decision stand"; the doctrine that decisions of the court constitute a precedent standing for future guidance, and that all subsequent decisions on the same issue should be bound by the precedent.

statute of limitations The statute which sets up the respective periods of time within which various actions or proceedings must be initiated. For example, a jurisdiction may have the following statute of limitations on the filing of a criminal complaint against the accused: no limitation for murder, embezzlement of public monies, or falsification of public documents; three years for a felony other than the above; one year for a misdemeanor.

statutory immunity Immunity from liability granted by statute. For example, despite occurrence of a contractual or tortious injury, recovery in a given case is not possible because statutory law prohibits recovery.

statutory law Law created by act of the legislature, in contrast to law created by judicial decision or by an administrative agency.

strict liability The concept whereby a manufacturer or distributor is held liable for any injury resulting from an unreasonably dangerous or defective product.

subrogation The substitution of one person for another, with reference to a lawful claim or right, so that the substitute acquires all rights of recovery.

suit See *lawsuit.*

tort A private injury or wrong, committed with or without force, to a person or his property for which a court will grant a remedy; a civil injury not arising out of contract.

trespasser A person who goes upon the premises of another without invitation, either express or implied, or whose presence there is not suffered or tolerated. It is not necessary that the trespasser, in making the unlawful entry, should have an unlawful intent.
ultra vires "Beyond the scope of power or authority.".
venue Refers to the county or judicial district within which a case can be tried. In most states, venue determined by statute.
verdict The jury's answer to the court concerning a factual issue given to it.
warrant An order made by a judicial officer, on behalf of the state, commanding a law enforcement officer to arrest a specified individual, search a specified location, or seize specified items.
writ An order issued by the court for the purpose of compelling a person to do or not to do something. See *mandamus*.

Index